Glimpses of VEGETABLE SCIENCE
for Competitive Examinations

THE AUTHORS

Dr. Anoj Yadav was born on 11 July 1995 in Azamgarh district of Uttar Pradesh. Dr. Yadav passed his Bachelor of Agriculture degree in first division from Tilakdhari Postgraduate College, Jaunpur, which is affiliated to Veer Bahadur Singh Purvanchal University, Jaunpur. He did his M.Sc. (Ag.) Horticulture degree in first division from Babasaheb Bhimrao Ambedkar University (A central University) Lucknow. Dr Yadav completed his Ph D. horticulture (Vegetable Science) Degree in first division from Acharya Narendra Deva University of Agriculture and Technology, Kumarganj, Ayodhya. Dr. Yadav is presently working as Technical Assistant in the Department of Agriculture in the government of Uttar Pradesh, India. The author has qualified NET conduct by ASRB of ICAR for three times i.e. during 2020 (Vegetable Science), 2021 (Fruit Science) and 2023 in Vegetable Science. The author has so far published 10 research papers, 20 book chapters and 60 popular articles in different reputed referred national and international journals and magazines. Dr. Yadav conferred with Young Scientist award and his Ph. D. thesis was recognized with best thesis award.

Dr. Gulab Chand Yadav, University and Vice- Chancellor Gold medalist in B.Sc. (Ag.) and M. Sc. (Ag.) from Acharya Narendra Deva University of Agriculture and Technology, Kumarganj, Ayodhya, respectively. Doctorate degree from Department of Genetics and Plant Breeding, G. B. Pant University of Agriculture and Technology, Pantnagar. Fellow of Indian Society of Vegetable Science, Varanasi and International Society for Noni Science, Chennai. Conferred ten different awards from different scientific societies of the countries and Best Teacher Award of ANDUAT Kumarganj Ayodhya, and more than 50 sport medals and NCC certificates. served as SRF for one year in department of Genetics and Plant Breeding, G. B. Pant University of Agriculture and Technology, Pantnagar. Has served in various capacity including Assistant Professor, Associate Professor, Professor and Head, Farm-in charge and teaching in charge in the Department of Vegetable Science, Acharya Narendra Deva University of Agriculture and Technology, Kumarganj, Ayodhya, CDI in Cane Department of UP, and presently working as Professor, Department of Horticulture, Babasaheb Bhimrao Ambedkar University (A central University) Lucknow. Also worked as external examiner/experts in various universities for setting of question papers, theses evaluation and viva –voce and conduct of practical exams. Developed five varieties, 3 OP varieties of bottle gourd (Narendra Prabha, Narendra Pooja, Narendra Kamna), 1 OP variety of brinjal (Narendra Suyog) and 1 hybrid variety of bottle gourd(Narendra Sita) during 19 years of service in Acharya Narendra Deva University of Agriculture and Technology, Kumarganj, Ayodhya. Published more than 160 research papers, 15 books more than 100 popular articles. Guided 61 post graduate students (43 Masters and 18 Ph. D.) and members in the advisory committee of more than 250 masters and doctorate students. Delivered 20 TV and 45 radio talks and lectures in more than 100 gosthies on different aspects of vegetables.

Glimpses of
VEGETABLE SCIENCE
for Competitive Examinations

by

Anoj Yadav

G. C. Yadav

2025

Daya Publishing House®

A Division of

Astral International Pvt. Ltd.

New Delhi – 110 002

ISBN: 978-93-5919-960-3

Published by : **Daya Publishing House®**
A Division of
Astral International Pvt. Ltd.
– ISO 9001:2015 Certified Company –
4736/23, Ansari Road, Darya Ganj
New Delhi-110 002
Ph. 011-43549197, 23278134
E-mail: info@astralint.com
Website: www.astralint.com

Preface

Vegetable science has become an important branch of horticulture in many parts of the world. Today in India vegetable production is being viewed as a high growth industry. Yet, the productivity of vegetables in our country is much low but there is great scope to increase it.

Agriculture research in India is well recognized as a highly demanding task to be dealt by highly knowledgeable and dynamic scientists. Agricultural Research Service under Indian Council of Agricultural Research has well design examination system for screening to ensure this inevitable quality in its scientists at various levels.

This book "**Glimpses of Vegetable Science for Competitive Examinations**" is design as a capsule of information on vegetable crops for the use of ARS, ICAR-SRF as well as entrance examinations of ICAR and SAU's aspirants . The information's are design in such a way that the maximum information could be pumped into the reader with lessor ready effort. Apart from major vegetable crops details on \minor vegetable crops tuber crops, spices and condiments, IPR, plant protection and other aspects are also furnished to ensure a complete and thorough reading. Model question papers given at the end of this book will certainly take the reader a long way ahead in competitive examinations.

We appreciate the efforts of M/s Astral International Pvt Ltd, New Delhi in executing a good job in time. The authors would welcome suggestions from readers to improve this book in its next edition.

Anoj Yadav

G. C. Yadav

Contents

CHAPTER 1

Introduction

- Vegetables are the herbaceous plant of which some portion is eaten either coocked or raw during the principal part of the meal.
- Study of vegetable cultivation is called olericulture
- India grows the largest number of vegetable crops in the world
- Father of vegetable L.H. Bailey
- Acc. To ICMR (Indian Cauncil of Medical Research) recommendation for balance diet:
    - ☆ 300g vegetable per capata per day
    - ☆ Out of 300g = 125g leafy vegetable + 100g root and tuber + 75g other vegetable
    - ☆ Per capita availability of vegetable=250g
- Vegetable crops in India occupy only 2.8% of the total cropped area
- Uttar Pradesh is the largest producer of vegetables in India.
- Dr. H.B. Singh is the pioneer vegetable breeder in the country.
- Onion (big) for export is grown in Nasik, Pune and Satara district in MH.
- Recently gherkin has become an important vegetable for export.
- Cauliflower does not require vernalization for flowering unlike cabbage
- In India, onion cultivars are short day type which require 10-11 hiurs of day length of flowering
- Cluster bean, hyacinth bean cowpea and watermelon are more tolerant to drought then other vegetable crops
- Permanent method of preservation of fruit and vegetable: Eclusion of moisture
- Vacum cooling is most suitable for leafy vegetable
- Leafy vegetable are rich source of Iron

- Fruit pH above 4.5 is not acceptable for processing industries where heating duration for sterilization are fixed
- Whenever the calcium content in tomato fruit fall below .2% BER is observed
- Arka Samrat and Arka Rakshak variety of tomato are resistant to TLCV, Bacterial wilt and Early blight (MDR)
- Succulometer is used for measuring the maturity of sweet corn
- Flame peeling: Onion and Garlic
- Moisture: Ash gourd/ wax gourd 96.5g/100g) highest.
- The optimum temperature for most of vegetables is b/w 10 to 24°C
- Often night temperature is more critical than day temperature
- Onion and radish become more pungent at high temperatures.
- Arka Gaurav and Arka Basant variety of capscicum show indeterminate fruiting
- Jawahar-218 variety of chilli developed by pedigree selection
- MDU-1 variety of chilli developed by mutation breeding
- Potassium present in plant in the solution of **cell sap**
- Major mineral present in fruit and vegetable is potassium
- Nano DAP cantain 8% N and 16% P_2O_5
- The soil moisture of most of vegetable crops ranges b/w 70-80 % of the availability in the root zone
- Highest area in Vegetable crop: Potato > Tomato > Onion > Brinjal
- Highest production in vegetable crop: Potato> Tomato > Onion > Brinjal
- Highest productivity in vegetables: Cassava
- Inflorescence of cabbage known as catken
- Winged bean a vegetable crop is all part are edible
- Edible portion of celery is leaf stalk
- Metromorphy is observed in cole crops, while patromarphy is observed in solanaceous crops
- Freezing invalves principle of crystallization, whereas Zero energy cool chambers invalve the principle of evaporative cooling

Vegetable Research Institutes

- Project Directorate of Vegetable Research (PDVR) was started in New Delhi, 1987
- H.Q. of PDVR shifted from New Delhi to Varanasi, 1992
- PDVR was upgraded as Indian institute of vegetable research (IIVR) Varanasi (UP) in 1999
- Central Tuber Crops Research Institute (CTCRI), Thiruvananthapuram, Kerala
- Central Potato Research Institute (CPRI), Shimla, Himachal
- National Horticultural Research and Development Foundation (NHRDF), Nasik, Maharashtra
- Directorate on Onion and Garlic Research (DOGR) Nasik Maharashtra (Previously known as NRC for Onion and Garlic NRCOG)
- Directorate of Mushroom Research (DMR) Solan, Himachal Pradesh
- Indian Society of Vegetable Science (ISVS),1973
- Vegetable science (half yearly) is published: ISVS

CHAPTER-2

Classification of Vegetables

(A) Botanical Classification

Sr. No.	Family	Common Name	Botanical Name	Origin	Edible part	2n
			(I) MONOCOTYLEDONEAE			
1	Amaryllidaceae (Alliaceae)	Onion	*Allium cepa*	Central Asia	Bulb	16
		Multiplier onion	*A. cepa var. aggregatum*	-	Small bulb	16
		Top/Tree onion	*A. cepa var. viviparum*	-	Root and bulbis	16
		Garlic	*A. sativum*	Central Asia	Clove	16
		Leek	*A. porum*	Central Asia	Stem and leave	2n=4x=32
		Welsh onion	*A. fistulosum*	-	Stem and leave	16
		Shallot	*A. ascolonicum*	-	Young bulb and green leave	16
		Chive	*A. schoenoprosum*	-	Enlarged stem and leave	16,24,32
		Kurrat	*A. kurrat*	-	Green leave	2n=4x=32
2	Araceae	Taro	*Colocasia esculenta*	Srilanka	Corms	28
		Eddoe type	*C. esculenta var. antiquorum*	-	Corm and cormel	2n=2x=24 2n=3x=42
		Dasheen type	*C. esculenta var. globulifera*	-	Corm and cormel	2n=2x=24 2n=3x=42
		Tannia	*Xanthosoma sagittifolium*	-	Corms	26
		Elephant foot yam	*Amorphophyllus campanulatus*	-	Corms	26
3	Dioscoraceae	Yam	*Dioscorea spp.*			30-80
		Greater yam	*Dioscorea alata*	Indo-burma	Tuber	2n=4x=40
		Lesser yam	*Dioscorea esculenta*	Indo-burma	Tuber	2n=4x=40
		White yam	*Dioscorea rotundata*	Indo-burma	Tuber	2n=4x=40

Sr. No.	Family	Common Name	Botanical Name	Origin	Edible part	2n
4	Poaceae (Graminae)	Sweet corn	*Zea mays var. rugosa/sacchrata*	-	Immature Kernel	20
5	Liliaceae	Asparagus	*Asparagus officinalis*	Europe, Asia	Spears (young shoots)	20
			(II) DICOTYLEDONAE			
6	Amaranthaceae	Amaranthus	*Amaranthus spp.*	India	Leaves, stem	32
7	Chenopod-iaceae	Beet root	*Beta vulgaris*	Mediterranean region	Root	18
		Palak	*Beta Vulgaris var. bengalensis*	Indo-china	Leaves	18
		Spinach	*Spinacea oleracea*	Iran	Leaves	12
		Pigweed	*Chenopodium album*	-	Leave and tender twigs	2n=4x=34
8	Compositae	Lettuce	*Lactuca sativa*	Mediterranean region	Leafy heads	18
		Globe artichoke	*Cynara scolymus*	Mediterranean region	Flower bud	34
		Jerusalem artichoke	*Helianthus tuberosus*	USA	Tuber	2n=6x =102
9	Convolvulaceae	Sweet Potato	*Ipomea batatas*	S. America	Tuber	2n=6x=90
10	Cruciferare (Brassicaceae)	Cabbage	*Brassica oleracea var. capitata*	Mediterranean region	Head	18
		Cauliflower	*Brassica oleracea var. botrytis*	-do-	Curd	18
		Brussels sprout	*Brassica oleracea var gemmifera*	-do-	Head	18
		Sprouting Broccoli	*Brassica oleracea var. italica*	-do-	Flower bud	18
		Knol-khol	*Brassica oleracea var gongylodes or Brassica caulorapa*	-do-	Knob (Stem)	18
		Kale	*Brassica oleracea var acephala*	-do-	Leaves & shoot	18
		Chinese Cabbage	*Brassica Chinensis*	China	Leafy heads	20
		Turnip	*Brassica rapa*	China, India	Root	20
		Raddish	*Raphanus sativus*	Europe	Root and leaves	18
		leaf mustard	*Brassica juncea var. cunefolia*	-	tender leaves	2n=4x=36

Sr. No.	Family	Common Name	Botanical Name	Origin	Edible part	2n
11	Cucurbitaceae	Cucumber	*Cucumis sativus*	India	Fruit	14
		Muskmelon	*Cucumis melo*	Tropical Africa	-do-	24
		Snapmelon (Phoot)	*Cucumis melo var. momordica*	-do-	-do-	24
		Long melon (Kakri)	*cucumis melo var. utillisimus*	-do-	-do-	24
		Gherkin	*Cucumis anguria*	-do-	-do-	24
		Watermelon	*Citrullus lanatus*	-do-	-do-	22
		Round melon	*Citrullus lanatus var. fistulosus*	-do-	-do-	22
		Pumpkin	*Cucurbita moschata*	Mexico	-do-	40
		Summer squash	*Cucurbita pepo*	Mexico	-do-	40
		Winter squash	*Cucurbita maxima*	Mexico	-do-	40
		Bottle gourd	*Lagenaria siceraria*	S. Africa	-do-	22
		Bitter gourd	*Momordica charantia*	Indo-burma	-do-	22
		Ridge gourd	*Luffa acutangula*	Asia	-do-	26
		Sponge gourd	*Luffa cylindrica*	Assam	-do-	26
		Pointed gourd	*Trichosanthus dioca*	India	-do-	22
		Snake Gourd	*Trichosanthus anguina*	India	-do-	22
		Wax (Ash) gourd	*Benincasa hispida*	Java, Japan	-do-	24
		Ivy gourd	*Coccinia indica*	India	-do-	24
		Chow chow	*Sechium edule*	Mexico	-do-	28
		Spine gourd	*Momordia Cochinchinesis*	-	-do-	28
12	Euphorbiaceae	Cassava	*Manihot esculanta*	Brazil	Tuberous root	36
		Chekurmanis	*Sauropus androgynous*	Indo-burma	Green leaves	-
13	Labitae	Chinese Potato	*Coleus pervriflorus*	Africa	Tuber	-

Sr. No.	Family	Common Name	Botanical Name	Origin	Edible part	2n
14	Leguminoceae	Peas	*Pisum sativum*	Central Asia	Tender seeds	14
		French Bean	*Phaseolus vulgaris*	Mexico	Pod, seed	22
		Cluster Bean	*Cyamopsis tetragonolobus*	India	Pod, seed	14
		Winged Bean	*Psophocarpus tetragonolobus*	Africa	Pod, seed	18
		Lima Bean	*Phaseolus lunatus*	Guate mala	Pod	-
		Dolichos bean	*Lablab purpureus*	India	Pod	22
		Cow pea	*Vigna unguiculata*	Africa	Pod	22
		Agathi	*Sesbania grandifloƨa*	-	Flower	24
		Fenugreek	*Trigonella focnu graceum.*	-	Leaves	16
		Winged Bean	*Psophocarpus tetragonolobus*	Africa	Pod	18
15	Malvaceae	Okra	*Abelmoschus esculantus*	Africa	Tender fruit	130
16	Moringaceae	Drumstick	*Moringa Oleifera*	Africa, India	Green pods	28
17	Solanaceae	Potato	*Solanum tuberosum*	South America	Stem, tuber	2n=4x=48
		Tomato	*Lycopersicon esculentum*	South America	Fruit	24
		Brinjal	*Solanum melongena*	India	Fruit	24
		Sweet pepper	*Capsicum annum*	Mexico	Fruit	24
		Chilli	*Capsicum annum*	Mexico	Fruit	24
18	Umbeliferae	Carrot	*Daucus carota*	Afganisthan	Root	18
19		Coriander	*Coriandrum sativum*	Mediterranean region	Leaves	22
		Celery	*Apium graveolens*	-	Leaves	22
20	Lauraceae	Bay leaf	*Paurus nobilis*	-	Leaves	-
21	Aizoaceae	New Zealand Spinach	*Tetragonia tetragonioidea*	Mediterranean region	Tender leave	-
22	Basillaceae	Indian spinach	*Basella rubra*	India	Fleshy stem and roots	24
23	Polygonaceae	Rhubarb	*Rheum rhapontium*	-	Thik leaf stalk	44
24	Rutaceae	Curry leaf	*Murraya koenigii*	-	Leave	18

(B) Acid Present in Vegetables

- **Citric acid**: Tomato, Beet root, Leafy vegetables, Legumes, Potato
- **Malic acid**: Carrot, Celery, Lettuce, Onion, Broccoli

(C) Aroma Containing Compounds

- **Cucumber:** Nonadienal

- **Cabbage**
  - ☆ **Raw -** Allyll isothiocynate
  - ☆ **Cooked-**Dimethyl disulphide
- **Potato:** Dimethyl pyrazine
- **Radish:** Isothiocynate
- **Onion:** Allyl propyl disulphide
- **Garlic:** Diallyl disulphide

(D) A/c to Respiration

- **Climacteric:** Musk melon, Watermelon, Tomato
- **Non climacteric**: Cucumber

(E) Temperature Requirement

Sr. No.	Optimum temperature	Crop
1	25-27°C	Okra, Chilli, Sweet Potato, Water Melon, Musk Melon
2	30-25°C	Tomato, Brinjal, Pepper, Cucumber, Pumpkin, Onion, Garlic
3	18-25°C	Cauliflower, Cabbage, Radish, Potato, Carrot, Lettuce

(F) A/c to Photoperiod

Long day plant	Short day plant	Day neutral crop
Potato, Onion, Cabbage, Cauliflower, Radish, Lettuce, Spinach, Palak, Turnip, Carrot, Beet	**Sweet Potato,** Indian spinach, Dolichos Bean, Cluster Bean, Winged Bean	Tomato, Brinjal, Chilli, Sweet pepper, Okra, Cucurbits, Amarnathus, French Bean, Cowpea

(G) Water Loss

Low/ Least	Tomato
Moderate	Sweet pepper, Cabbage
High	Summer squash
Very high	French bean leafy green

(H) Isolation Distance

Crop	Foundation Seed (m)	Certified Seed (m)
Potato	5	5
Tomato, Fenugreek, Lettuce	50	25
Brinjal	200	100

Crop	Foundation Seed (m)	Certified Seed (m)
Chilli, Sweet pepper,Okra	400	200
Pea	20	10
Cowpea, french bean	10	10
Cole crop, Radish,Turnip, Beet, Spinach	1600	1000
Cucurbits	1000	500
Carrot	1000	800
Onion	1000	500
Bulb Production (Onion)	5	5

(I) Pollination Mechanism

Type	Extent of cross pollination	Crops
Self pollination	<5%	Tomato, Lettuce, pea
Often cross pollination	5-30%	Chilli, Capsicum, Brinjal, Okra
Cross pollination	>30%	All cole crop and Cucurbits

(J) Dichogamy

Protoandry	Protogyny
Carrot	Cole Crop
Onion	Okra
Beet	Chiili
Parsley	EFY
Leek	

(K) Self Incompatability (SI)

- **Sporophytic SI –** Cole Crops, Radish, Turnip, Beet
- **Gametophytic SI -** Tomato, Sweet Potato

(L) Dioecious

- Pointed Gourd, Scarlet Gourd, Spine gourd, Asparagus, Spinach, Yams, Beet root

(M) Tolerance to Soil Acidity

Less tolerant (pH-6.8-6)	Moderate (pH 6.8-5.5)	Highly tolerant (pH 6.8-5)
Cole crops, Okra, Musk Melon	Brinjal, Tomato, Chilli	Potato, Sweet Potato, Water Melon

(N) Tolerance to Soil Salinity

Less tolerant	Moderate	Highly tolerant
Potato, Sweet Potato Brinjal, Radish Pea	Chilli, Cole crops Tomato, Onion	Bitter Gourd, Ash Gourd, Palak Turnip, Lettuce, Beet

(O) Rooting Depth

Very Shallow (15-30 cm)	Very Deep (120-180 cm)	Moderately Deep (80-100 cm)
1. Onion 2. Lettuce 3. Small radish	1. Sweet Potato 2. Water Melon 3. Pumpkin 4. Tomato 5. Asparagus 6. Cluster bean	1. Brinjal 2. Chilli 3. Musk Melon 4. Turnip

(P) Respiration Rate of Produce

Very low	Low	Moderate	High	Very High
Potato Onion	Sweet Potato Cabbage	Chilli Tomato Carrot	Peas Beans	Leafy greens, Green Onion, Musk Melon, Water Melon, Cauliflower, Asparagus

(Q) Largest Producer in the World

Country	Crops
USA	Sprouting Broccoli
India	Cauliflower, Chilli, Okra
China	Garlic, Capsicum, Sweet Potato, Brinjal
Nigeria	Cassava
Africa	Yams

(R) Toxic Substances Found in Vegetable Crop

Toxic Substance	Crop	Toxic Substance	Crop
Apiin	Celery	Sinegrin	Cole crops
Ca Oxalates	Colocasia,EFY	Solasodine	Brinjal
Saponine	Tomato, Spinach, Asparagus	CN Glycosides	Cassava (Linuron-major toxic)
Cucurbitacins	Cucurbits	Haemaglutine	French Bean
Dioscorine	Yams	Oxalic acid	Amaranthus

Toxic Substance	Crop	Toxic Substance	Crop
Trypsin inhibitor	Peas and Beans, Sweet potato	Serotonin	Water melon
Choline estarase	Pumpkin	Tomatine	Tomato
Vicine and covicine	Broad bean		

(S) Type of Inflorescence in Vegetable Crops

Type of inflorescence	Crops
Raceme	Cole crops, cucurbits, radish
Cyme	Spinach, sweet potato, potato, broccoli, tomato, brinjal
Panicle	Palak, moringa
Spike	Beet root
Compound umbel	Carrot, coriander
Capitulum	Lettuce

(T) Common Name of Different Vegetables

Vegetable crops	Common Name
Tomato	Wolf Apple, Vilayati Baigan, Love Apple
Brinjal	Egg Plant
Lima Bean	Baby Potato, Double Bean, Butter Bean
Dolichos Bean	Indian Bean, Hycainth Bean
French Bean	Kidney Bean, Haricot Bean, Snap Bean, Navy Bean
Broad Bean	Horse Bean, Faba Bean
Winged Bean	Goa Bean, Four angled Bean, Vegetable of 20th century / wonder vegetable
Yam Bean	Potato Bean, Misrikand
Bitter gourd	Bitter cucumber, Balsam Pear
Summer squash	Bush Squash, Chappan kaddu, Field pumpkin,Vilavati kaddu
Pumpkin	Butternut squash, Vegetable of immense value
Ash gourd	White gourd, Wax gourd
Cassava	Tapioca, Mardioca, Yuca, Manioc
Drumstick	Horse radish tree, Muringa
Chekurmanis	Multivitamin Greens, Madura kheera, Vegetable of 21st century
Cylon spinach	Water leaf
Capsicum	Bell pepper, Sweet pepper
Cowpea	Black eyed pea, Southern Pea
Musk Melon	Wholesome food
Spinach beet (Palak)	Swiss chord, Garden Beet Beetleaf

(U) Types of Vegetable Gardens

- Home/Kitchen/Nutritional garden - Area required for home garden 250 square meters
- Market garden:
  - ☆ Mostly situated near big cities
  - ☆ Intensive cultivation is followed
  - ☆ Many crops are grown simultaneously
  - ☆ 10-20 km Periphery market
- Truck garden - Extensive cultivation is followed one or two crops are grown.
- Processing vegetable garden:
  - ☆ Specialized cultivation practices are followed.
  - ☆ Mainly grown on contract basis.
- Vegetable forcing or control environment agriculture:
  - ☆ Intensive cultivation is followed.
  - ☆ Very small area is used.
  - ☆ Tomato and cucumber most suitable for vegetable forcing
- Vegetable garden for seed production cultivation:
  - ☆ Special types of vegetables are grown.
  - ☆ Contract cultivation
- Floating vegetable garden:
  - ☆ Mainly found in Jammu and Kashmir (Dal lake)
  - ☆ Shallow rooted vegetables are grown.
- Roof garden - Fresh leafy vegetables are grown.
  - ☆ Acc. to Thompson and Kelly 1979 (7 Type) of vegetable garden

(V) Inbreeding Depression

- **Very high depression:** Inbreeding should not be practiced for > 2 generations.

  Carrot

- **High depression:** Inbreeding can be practiced for 2-3 generations
  Onion
- **Moderate high depression:** Inbreeding can be practiced for 3-4 generations.
  Cabbage, Group I and II cauliflower, Broccoli, Brussels sprouts, Turnip, Sweet corn.
- **Low depression:** Inbreeding can be practiced for 4-5 generations.
  Snowball group of cauliflower (Group-IV Radish
- **Insignificant depression:**
  All cucurbits.
- **Lack of Inbreeding depression:**
  Self pollinated crops

(W). Varieties Developed by Mutation Breeding

Crops	Variety	Mutant type
Bitter gourd	MDU-1	Gamma ray mutant of MC-103
Ridge gourd	PKM-1	
Palak	Jobner	Green Spontaneous mutant from local type.
Tomato	S-12	X-ray mutant of Sioux.
	Maruthan (Co-3)	Mutant of Co-1
	PKM-1	Mutant of Annagi.
	Pusa Lal Meeruti.	Gamma ray mutant of K-1.
Chilli	MDU-1	Gamma ray mutant of K-1.
French bean	Pusa Parvati	X-Ray mutant of Wax Pod
Hyacinth bean	Co-10	Gamma ray mutant of Co-6
Okra	EMS-8	EMS-treated mutant of Pusa Sawani.
	MDU-1	

(X) Ancestors of Various Vegetable Crops

Pea	*Pisum elatius*
French bean	*Phaseolus aborigineus*
Radish	*Raphanus raphanistrum*
Beet	*Beta vulgaris spp. maritima*
Cluster bean (Guar)	*Cyamposis senegalensis*
Sweet potato	*Ipomea trifida*
Tomato	*Lycopersicon esculentum var. cerasiforme.*
Cucumber	*Cucumis hardwickii*
Watermelon	*Citrullus colocynthoides*
Garlic	*Allium longicuspis*

Onion and Shallot	*A. vavilovi*
Leeks	*Allium ampeloprasum*
Welsh onion/Japanese bunching onion	*Allium altaicum*
Globe Artichoke	*Cynara carcundulus (wild cardon)*
Cowpea	*Vigna unguiculata var. mensensii*
Lettuce	*Lactuca serriola*
Bell pepper	*Capsicum annum var. minmum*
Okra	*Abelmoschus tuberculatus*

(Y) Diameter During Harvest of Different Vegetables

Carrot - 2-4 cm (root)

Beet - 3-5 cm (root)

Turnip - 5-7.5 cm (root)

Knolkhol - 5-8 cm (root)

(Z) Different Product Produced from Different Vegetables

Product	**From Vegetable**
Saurkraut	Cabbage
Yerusseri	Immature Pumpkin
Fufu	Yam
Tuty-fruite	Bottle gourd
Petha	Ash gourd/ wax gourd
Acidic product Poi	Taro
Sago/ foufou	Cassava
Weikang	Sweet potato
Pachranga, Satranga	Cauliflower

Nutritional Value

Nutrient	**Source**
Carbohydrates	Tapioca (38.1%)
	Sweet potato (28.2%)
	Potato (22.6%)
Fibre	Chilli (6.8%)
	Green pea (4.0%)
	Cluster bean (3.2)
Protein	Lima bean (8.3%)
	Green pea (7.3%)
	Cowpea (4.3%)

Nutrient	Source
Fat	Potato (1.4%)
Vitamin -A	Ragi (14190μg/100)
	Black colocasia leaves (12000μg/100)
	Green colocasia leaves (10278μg/100)
Vitamin –B_1	Chilli (0.55mg/100)
Vitamin –C	Parseley (281mg/100)
	Drumstick leaves (220mg/100)
Calcium	Agathi (1130mg/100)
Phosphorus	Amaranthus (800mg/100)
Iron	Amaranthus (22.9mg/100)
Iodine	Okra, Onion

Vegetables are rich source of vitamin and mineral is known as "protective food"

- ☆ Carbohydrates serve as the chief source of energy food also called protein sparing food

Populations and their genetic constitutions

- ❐ Homozygous homogeneous. *e.g*- Purelines, Inbred, self-pollinated crops
- ❐ Heterozygous and heterogeneous *e.g*-Synthetic and Composite variety, cros-pollinated crops
- ❐ Homozygous and heterogeneous *e.g*- Multilines, mass selected populations.
- ❐ Heterozygous and homogeneous *e.g*- hybrid, clone

Genetic terms and their genetic constitutions -

Simplex - Aaaa

Duplex - AAaa

Triplex - AAAa

Quadriplex - AAAA

Nulliplex - aaaa

Seed treatment chemicals and agents in vegetable crops

Fungicide - Thiram

Bactericide - Streptocycline

Antagonistic fungi - Trichoderma spp

Insecticide - Carbofuron-60g/kg seeds

- Imidaclorpid - 10g/kg seeds.

Number of Stamens in different vegetable crops

No of Stamen	Crop
6	Cole crops, Radish, Yam,
5	Tomato,Brinjal, Carrot, Beet, Potato,Sweet potato
10	Cassava
2-5	Amaranthus

Seed rate of different vegetable crops

Crops	Seed rate/ha
Tomato	300-400g
	Hybrid – 125-175g
Brinjal	200-300g
Chilli	1-2kg
Cauliflower, Cabbage,	Early 500-600g
	Late – 350-400g
Knolkhol	800-1000g
Kale	350-400g
Carrot	8-10kg
Radish	9-12 kg
Beet	7-9kg
Turnip	3-4kg
Onion	8-10kg
Garlic	350-500kg clove
Leek	5-7kg
Pea	Early 100-120kg, Mid and Late 80-90kg
French bean	Bush type 80-90kg and Pole type 25-30kg
Cowpea	15-20kg
Cluster bean	15-40kg
Okra	Spring 18-20kg, Kharif 8-10kg
Palak	Winter 10-15kg and Summer -25-30kg
Spinach	37-45kg
Amaranthus spp.	2-2.5 kg
Fenugreek	Common methi -25kg and Kasuri methi 20kg
basella	12-16kg

Pollination

- **Anemophily-** Spinach, Beetroot, Palak, Amaranthus
- **Entomophily**- Onion, Carrot, Radish, Cole crop
  - ☆ **Feachers of flowers**
    - **Cleistogamy** – Lettuce

- **Chasmogamy**- Tomato, Capsicum
- **Protandary**- Onion, Carrot
- **Protogyny**- Cole crop, Okra
- **Hercogamy**- Lima bean

Classification based on the plant parts used as vegetable

- **Flower-** Agathi, male flowers of pumpkin
- **Prefloral apical meristem-** Cauliflower
- **Flower head-** Broccoli, globe artichoke
- **Modified above ground stem-** Knolkhol, asparagus
- **Modified stem-** Potato, Jerusalem artichoke, yam, elephant foot yam, taro, onion, garlic
- **Modified root-** Radish, carrot, beetroot, turnip, sweet potato
- **Modified tap root-** Chinese artichoke
- **Modified adventitious root-** Chinese potato
- **Fruits-** Brinjal, tomato, chilli, peas and beans, all cucurbits, okra
- **Corm-** Colacasia, elephant foot yam

Edible portion of vegetable crops

- **Placentae-** Cucumber, watermelon
- **Endocarp-** Ridge gourd, sponge gourd, ash gourd
- **Mesocarp and pericarp**- Pumpkin and musk melon

Useful compounds present in vegetables

Vegetables	Compounds	Role in disease prevention
Onion, garlic	Sulphur compounds	Reducing blood cholesterol
Cole crops	Indoles and dithiolthiones	Prevention of cancer
Bitter gourd	Cheratin	Effective against diabetes
Onion	Diphenyl amine	Effective against diabetes
Celery	3-n-butyl pthalide	Effective against hypertension
Yams	Diosgenin	Manufacture of contraceptive drugs
Onion, garlic	Quercetin (Bioflavonoid)	Protection against cancer & heart disease
Lettuce	Lutein	Nutritional antioxidant
Onion	Allicin	Antibacterial activity
Cabbage	Indole-3-carbinol	Against bowl cancer
Broccoli	Sulphoraphane	Anticancer activity
French bean, broccoli	Kaemferol	Anti cancer and cardiovascular disease
Brinjal	Nasunin	Anticancer activity

Male sterility systems in vegetable crops

Types of male sterility	Inheritance pattern	Vegetable crops
1. Pollen sterility		
CMS	Not occur in nature	
GMS	Singe recessive gene (ms)	Tomato ,brinjal, chilli cucurbit, cole crops
	Duplicate dominant gene	Cauliflower
CGMS	Singe recessive gene (ms)	Onion, radish, sweet pepper, cole crops and turnip
	Two recessive gene	Beet
	Single dominant gene	Carrot
2. Staminal male sterility *(sl)*	Single recessive gene	Tomato
3. Structural/Functional male sterility		
i. Positional sterility *(ps)*	Single recessive gene	Tomato, brinjal, sweet pepper
ii. Exserted stigma sterility *(ex)*	Single recessive gene	Tomato

A. Genetic Male Sterility (GMS) or Nuclear Male Sterility (NMS)

Crops	Gene	Commercially utilised	Variety
Tomato	Single recessive gene	*ps-2 gene*	-
Chilli	-do-	*ms-12 & ms-3 gene*	CH-1, CH-3
Muskmelon	-do-	*ms-1 gene*	Punjab Hybrid-1

B. Cytoplasmic Genetic Male Sterility (CGMS) in Vegetable Crops

Crops	Gene	Commercially utilised	Variety
Chilli	Single recessive gene	*ms-2*	Arka Meghana, Arka Sweta, Arka Harita, Kashi Surkh
Onion	-do-		Arka Kirtiman, Arka Lalima
Carrot	-do-		Pusa Nayanjyothi, Pusa Vasuda

C. Cytoplasmic Male Sterility (GCMS) in Vegetable Crops

Crops	Types	Commercially utilised
Onion	*S and T cytoplasm*	S-cytoplasm type
Carrot	*Peataloid (pr), brown anther (ba) and gummifer (gu)*	Petaloid type
Cauliflower, Cabbage, Brussels sprout, Sprouting Broccoli	-	Ogura cytoplasm

❖ Derivation of Vegetable for Plant Tissue

- **Swollen inflorescence-** Broccoli
- **Stem sprout-** Asparagus

- **Main bud-** Lettuce
- **Axillary bud-** Brussels sprout
- **Petiole-** Celery
- **Swollen leaf base-** Leek
- **Leaf blade-** Spinach
- **Swollen taproot-** Carrot
- **Stem tuber-** Potato
- **Swollen hypocotyl-** Beetroot
- **Bulb (Underground bud)-**Onion
- **Flower bud-** Artichoke
- **Placental intralocular tissue (septum)-** Tomato

Effects of Ethylene in Vegetables

Crops	Effect
Lettuce	Russet sprouting, discolouration
Carrot, beet	Bitterness
Celery	Blanching
Cabbage	Leaf abscission, undesirable flavour
Asparagus	Fibrousness
Cucumber	Degreening
Peas	Toxin pisatin formation
Sweet potato	Hard core, production of phenolics
Tomato	Increased lycopene and carotene

Edible Colour Rich Varieties In India

Crop	Variety	Pigments
Carrot	Pusa Asita (Black)	Anthocyanin
	Pusa Rudhira	Lycopene
	Pusa Nayanjyothi	ß - carotene
Radish	Palam Hriday	Anthocyanin
Amaranth	Pusa Lal Chaulai	Anthocyanin
Broccoli	Palam Vichitra	Anthocyanin
Tomato	**Pusa Rohini**	Lycopene
Beetroot	Detroit Dark Red	Anthocyanin
Paprika	KTPL-19	Capsanthin
Pumpkin	Arka Chandan	ß -carotene

CHAPTER-3

Solanaceous Vegetable Crops

Tomato

Common Name : Wolf Apple, Vilayati Baigan:

Botanical Name : *Lycopersicon esculentum* Mill. 2n=2x=24:

Origin : South America (Peru –Ecuador –Bolivia)

- Self pollination
- Type of fruit: Berry (multi seeded).
- Edible part: Fruits
- TGRC: Tomato Genetic Resource Centre California, USA.
- CGR: Centre for Genetic Resource, Netherland.
- Red colour in tomato is due to lycopene (20-50mg/100g).
- Lycopene is highest at 21-24^{0}C temperature and this pigment drops off >27^{0}C
- Above the 30^{0}C red colour suppressed, if temp. <30 red colour again develop
- Above 40^{0}C – lycopene destroyed and no red colour develops.
- Lycopene has also been reported to enhance resistance against body X-rays irradiation.
- During ripening the chloroplast are transformed into chromoplast.
- Yellow – carotenoid pigment (β carotene).
- Tangerine – prolycopene pigment.
- Fruit colour of tomato is determined by colorimeter.
- Major constraint to low fruit set in tomato: Low temperature and poor anther dehiscence
- Tomato variety Severianin shows better fruit set at low temperature due to parthenocarpy
- Popular variety of tomato: Pusa Ruby and Arka Vikas

- Pusa Divya only hybrid developed by utilizing the male sterile in india
- First hybrid variety of tomato: Karnatka, 1972
- For low temperature variety: Pusa Sheetal (grown in 8°C)
- For high temperature variety: Pusa Hybrid-1 (grown in 41°C)
- For low and high temperature variety: Pusa Sadabahar
- Extremly early variety: Hissar Arun and HS-102
- Pear shaped fruit variety: Pusa Sadabahar, Punjab Chhuhara, Roma
- Variety suited for long distance: Punjab Chhuhara
- Seed rate
    - ☆ For normal- 350- 400g/ha
    - ☆ For hybrid – 100-150g/ha
- Major pest of tomato - fruit borer (*Helicoverpa armigera*)
- Tomato is susceptible to Frost
- Tomato fruit is generally regarded as a protective fruit
- In adverse condition tomato crop is sprayed with PCPA (para-chlorophenoxy acetic acid) for higher fruit setting
- Seed treatment with PCPA 10-20ppm improved fruit set in tomato.
- Punjab Chhuhara (pear shaped) is determinate LCV susceptible variety of tomato
- Tomato is globally popular and widely grown vegetable crops.
- It is a whole are partly used in maximum number of dishes prepared globally for daily use, therefore, it may consider as a 'Queen of vegetables'.
- It is known for its outstanding nutritive value.
- The first known record of tomato is in the year 1554.
- Miller (1768) gave the Name L. *esculentum* from *Solanum lycopersicum*. The genus *Lycopersicon* differs from the *solanum* by the absence of spine in the pinnatified leaves and narrow anther tips, which exhibit longtudinal dehiscence But, again Peralata and spooner give the Name *Solanum lycopersicum* L. in year 2000.

Muller 1940 divided into two sub-genera *viz.*

Eulycopersicon, (red fruit) containing carotenoid and lycopene pigment	**Eriopersicon,** green fruited species with anthocyanin pigmentation
L. *esculentum*, L. *pimpinellifolium* and L. *cheesmanii*	L. *peruvianum* L. *chilense*, L. *hirsutum*, L. *parviflorum*.

- Rick in 1979 and Tayler in 1986 divided lycopersicon into 9 species.
- Stemans are 5, anther bright yellow.
- Tomato produces perfect, hermaphrodites and pentamerous flowers.

- The typical red colour of ripe tomatoes is mainly due to lycopene
- 19 different types of carotene are present in red tomato.
- Yellow colour of tomato is due to β-carotene.
- Tomato seeds contain 24% oil, which is extracted from the pulp and residues and are used in canning industry.
- The semi-drying oil is used as salad oil and in the manufacture of margarine.
- Tomato contains steroidal glycoalkaloid known as α-tomatin ($C_{50} H_{83} NO_{21}$) (found in leaves, flower and fruits)
- Small hard tomato fruit contains tomatins about 300mg/100g of edible portion and in green fruit about 150mg/100g.
- It also contains small quantity of solanine (5mg/100g)
- Citric acid is the major constituent (.43%) then malic acid (.08%).
- The optimum temperature in the range of 18-24^{0}C for growth, flowering, fruiting and fruit ripening.
- When mean temperatures are above 40^{0}C, the fruit remain green because the chlorophyll degrading mechanism is inactivated.
- About 200-300g seed are needed to raise seedling in about 100m^2 areas for transplanting in one hectare.
- At closere spacing, early and total yields are higher but fruit size is reduced and incidence of insect pest and diseases are more.
- There are **five classes of tomato fruit maturity** as follows

1. Mature green stage	-
2. Breaker/turning stage	¼ blossom end of the fruit show pink colour, Suitable for long distance transport
3. Pink stage	3/4 blossom end are pink, Suitable for local market
4.Ripe stage	Table/salad purpose
5. Over ripe	Processing

Important species of tomato

Scientific Name	**Feature**
Solanum lycopersicon var. cerasiformae	Cherry tomato/ancestor of cultivated tomato
L. pimpenellifolium (Current tomato) SP+CP. (self and Pollination and cross Pollination)	Resistance to Fusarium wilt, early blight, leaf curl, bacterial speck and low temperature tolerance and high ascorbic acid content
L. cheesmanii	Resistance to salinity and high TSS

Scientific Name	Feature
L. peruvianum (cross pollination)	Resistance to leaf curl virus, TMV root knot nematode and rich source of ascorbic acid
L. hirsutum (self-fertile, self incompatible)	Excellent source of resistant to TMV, bacterial canker and bacterial speck
L. chemieleskii	Rich sourse of TSS and it is **cross pollinated** species
L. pennellii	Drought tolerance
Solanum lycopersicoides	Cold tolerance.

- *L. peruvianum and L. pimpenellifolium* are good source of ascorbic acid.
- Interspecific crosses between *L. esculantum* and *L. pimpinellifolium* are easily made.
- Embryo abortion occurs between *L. esculantum* and *L. peruvianum* and this can be overcome through embryo rescue technique.
- Of the 12 chromosome pairs, 1 is the largest chromosome, chromosome 2 is a nucleolus organizing (satellite) chromosome and 12 is the shortest chromosome.
- C.M. Rick Tomato Genetics Resource Center (TGRC), located at the University of California at Davis, USA.
- AVRDC, Taiwan (China) has the largest collection of tomato accessions.
- Hybrid seed production seed parent (female) and pollen parent (male) are grown in the ratio of 10-12:1.
- Pollen parents are generally sown 1 -2 weeks earlier than the the female parents to provide sufficient pollen.
- J- shaped glass tubes are most useful for direct transfer of pollen to the stigma.
- The potential sterile line known is ms10.
- **Cultivars** developed through mutation breeding
  - ☆ S-12: X-ray mutant of Sioux
  - ☆ Marusthal (CO-3). EMS mutant of CO-1
  - ☆ PKM-12: Mutant of Annagi
  - ☆ Pusa Lal Meeruti Gamma ray mutant of Meeruti
- Flat Large Red, Red Cherry, San Marzano variety are resistant to buck eye fruit rot (*Phytophthora paradisitica*)
- Tomato leaf curl Gemini virus (TLCV) is a very severe disease and transmitted by vector whitefly (*Bemisia tabaci*).

- Fruit borer (*Heliothis armigera*) and whitefly (*Bemisia tabaci*) are the important pest of tomato.
- There are presence of granular trichomes, tomatine and protinage inhibitor in wild species.
- Severity of nematode infestation is more in sandy soil due to fast development of nematode population and galls.

Varieties of Tomato

Varieties	Breeding method	Feature
Pusa Red Plum	*S. lycopersicon × S. pimpenellifolium*	Rich in Vit-C
Pusa Ruby	Sioux × Improved Meeruti	Most famous variety
Pusa Early Dwarf	Improved Meeruti × Red Cloud	-
Pusa -120		Resistant to RKN
Pusa Rohini and Pusa Gaurav		Suitable for long transportation and processing
Arka Meghali	Arka Vikash × IIHR-554	
Hissar Anmol	Hisar Arun × *S. hirsutum*	Resistant to leaf curl virus
Punjab Chhuhara	EC-55005 × Punjab Tropics	Suitable for long transport

Tomato plant has two types of growth habits

Compact or Determinate	Indeterminate or tall spreading
Inflorence occur almost at every internote	Inflorescence occur at every third inter notes.
Self topping or self pruning type	Single stem training and main axis continues growing indefinite indefinitely.
Papular in North India	Popular in South India and Hills
Plants have short internote both height and breadth is limited.	vigorous growth usually with long internodes
Grown in field	Grown in Polyhouse
Variety- Pusa Ruby, Pusa Divya, Arka Vikash, Arka Saurabh, Arka Ahuti, Arka Abha, Pant Bahar, Pant T-3, Best of All, Sioux, Narendra Tomato-1, NDT-5	Pusa Early Dwarf, Pusa Gaurav, Pusa Sheetal, Pusa Uphar, Pusa Sadabahar, KT-10, KT-12, Punjab Chhuhara, Punjab Kesari, Sweet-72, Narendra Tomato-2, NDT-120

Growth is continued by slow initiation or appearance of flower.

- For processing purpose of tomato Brix in the range of 4 to 6 and pH less than 4.4.
- It has been usually observed than indeterminate forms have more soluble solid then the determinant forms.
- Interspecific hybridization has been used extensively in tomato due to its narrow genetic basis.

- A great number of molecular markers (RFLP, AFLP, RAPD, CAPs, SSR) are mapped onto the genetic map of tomato.
- A period of drought follower why sudden heavy irrigation during fruiting may cause cracking of fruit
- Chemical used for tomato sauce preparation sodium benzoate
- The family solanaceae is divided into two subfamilies, Solanoideae and Cestroideae.
- *L. chilense and L. peruvianum* highly polymorphic species.
- It has a little exserted stigma with mostly inbreeding but slight out breeding also and self compatible.
- *Solanum pennellii,* now known as *Lycopersicon pennellii* is closest to the cultivated tomato.
- Both L. *esculantum and L. esculantum* var. *cerasiforme* can be intercrosed.
- Tomato was introduced in India during British period in the year 1828 by the Royal Agri- Horticultural society, Calcutta.
- In India, the first F_1 hybrid of tomato "Karnataka" was released for commercial cultivation by IAHS Bangalore in the year 1973.
- The species, *pimpinellifolium, peruvianum, hirsutum, and chilense* have been used widely as donors of genes for disease resistance in tomato breeding.
- Pusa Uphar is highly tolerant to fruit borer.
- The tomato leaf curl virus diseases occurring in Northern and Southern regions of India perhaps have diverse origin of the viruses.
- The Bt genes (GUi Ai, G42Aa (cry AB) are being utilised to develop transgenics having resistant to fruit borer.
- A gene HQT of the antioxidant chlorogenic Acid (CGA) which protects the tomato against the attack of bacterial diseases and stress.
- Genetic mapping and QTL (quantitative trait loci) analysis by RFLP technology have been reported in tomato.
- Spray of calcium tried to gram per litre of water is effective to overcome calcium deficiency and in case of Boron deficiency soil application of borax 20 to 30 kg per hectare is recommended.

Brinjal

Brinjal or egg plant synonyms Eggplant, Garden egg guinea squash

Scientific Name - *Solanum melongena* L.

Origin - India

2n=2x= 24.

- India is considered as the centre of origin of brinjal.
- The brinjal was first used in India.
- Satisfactory seed yield is approximately 150-200kg/ha.
- The 1,000 seed weight is approximately 5g but smaller seeded cultivars have 1000 seed weight of about 4g.
- Pigment present in Brinjal is anthocyanin.
- Brinjal fruit are good source of vitamin B.
- Brinjal is rich in chlorine
- Gene dominance in fruit colour in eggplant:Purple>Green>White
- Gene dominance in fruit shape: Elongated>Round>Oval
- Round fruits are baked or boiled for the preparation of smashed product called **Bharta (***BH-2 and Kashi Brinjal Green Round* variety suitable for bharta making**).**
- Pusa purple long (extra early maturity).
- Pant Samrat (Phomopsis blight and bacterial wilt resistant).
- Black beauty (nematode resistant)
- Small round fruits are used in curry, especially in Southern India.
- Phenols are responsible for discoloration, therefore less phenol contain is considered good for processing.
- Angel fruits are used for cure of toothache.
- It has also recommended for those suffering from liver complaints.
- The white brinjal is said to be beneficial for diabetic patients.
- The green leaf of brinjal plant are source of appetizer, aphrodisiac, cardio tonic beneficial in Vata and Kaph and contain antioxidant *i.e.* Vitamin C.
- The de-chlolestrolizing action is attributed to the presence of polyunsaturated-fatty acids (linoleic acid) which are present in flesh and seeds of the fruit in higher quantity (65.1%).
- The species, *S. aethiopicum and S. macrocarpon* are cultivated for the fruits as well as for their edible leaves.

- A wild type of brinjal with deeply variegated leaves (*S. sisymbrifolium* Lam.) is considered the ancestral species of *S. melongena*.
- Brinjal is grown as herbaceous annual while it is perennial in the tropics.
- Inflorescence is cymose.
- Flowering occurs usually solitary but sometimes it constitutes a cluster of 2-5 flowers.
- There are 4 flower types in brinjal called heterostyly *viz.,* (a) Long styled, (fruiting 70-86%), (b) Medium styled (fruiting 12.5-55.6%) (c) Short styled , and Pseudo styled in which no fruiting takes place.
- Short style and pseudo short styled flowers are reported to be male sterile.
- Brinjal is considered as often cross pollinated crop, natural cross pollination varies from 0.7 to 15%. out-crossing is performed by insect.
- Purple colour varieties are much in demand.
- The purple colour and its various shades in brinjal are due to the anthocyanin pigment.
- Green colour in brinjal due to chlorophyll.
- White colour brinjal is mostly used in summer months there is very less bitterness and fibreness.
- The green fruited brinjal is more preferred in Odisha, Karnataka, Bihar and West Bengal
- In North India, purple or violet and black colour varieties are preferred.
- In South India, small round or white stripped brinjal is preferred.
- The long green brinjal are prefered in **Karnataka and Bihar.**
- Optimum range of temperature for seed germination and vegetative growth is 25-30^{0}C, while temperature range of 25^{0}C is ideal for shooting fruiting , adversely affected if temperature exceeds the above 35^{0}C.

Seed rate

- For normal 300g
- For hybrid 150g

Sowing time and transplanting

- Northern plains
  - ☆ Autumn-winter crop seeds are sown in June- July and seedlings are transplanted in July -August
  - ☆ Spring-summer crop seed are sown in Nov- Dec. and transplanting during Feb- March

☆ Rainy season crop seed are sown in March and seedling are transplanted in April.

- In South India, the crop can be grown throughout the year
- In the hilly areas, the seedlings are raised in April and transplanted in may.
- Lack of surface glossiness is another indication of maturity.
- Fully matured or over matured fruit loss their shiny, glossy appearance and change to greenish bronze or yellow in colour.
- Delayed removal or failure to harvest horticultural mature fruit reduces flowering and subsequent fruit production.
- Brinjal fruit become ready to harvest after 16-17 days after anthesis in long fruited.
- 25-30 days after anthesis in oblong fruited and 35 -45 days in round fruited varieties.
- In general, 60-90 days are required from transplanting to first harvest.
- Fruit can be stored in good marketable condition for 7 to 10 days at 10 to 15^{0}C and 95% RH
- Brinjal fruit are sensitive to chilling and storage below 10^{0}C caused chilling injury, often exhibited a surface pitting and decay.
- Nagia and Kida were the first researchers who observed hybrid vigour in brinjal during 1926.
- Phomopsis fruit rot is recessive and polygenic in nature and the resistance source of wild species like *S. xanthocarpum, S. indicum, S. gilo, S. khasianum and S. sisymbriifolium.*
- Shoot and fruit borer is a serious insect pest of brinjal and the resistance source are available in wild-types such as *S. gilo, S. integrifolium S. xanthocarpum and S. sisymbrifolium.*
- Genera Solanum, Capsicum, and Lycopersicon are intergraftable.
- Side grafting and patch budding have been found successful.
- Glycoalkaloids; Alpha-solasonin and solamargin are linked with flavour quality.
- Species for leaves purpose *S. blumei, duplisinuatum, nodiflorum*
- Species for Leaves and fruits - *S. aethiopicum, gilo, incanum, macrocarpon, nigrum, Americanism.*
- Brinjal is a native of India and perhaps China is a secondary centre of origin.
- *S. incanum* is a progenitor of the cultivated species, *S. melongena* as reported by others earlier.

- Both *S. incanum and S. melongena* var. *insanum* are crossable with *S. melongena* and produce fertile hybrids.
- Fruits are born solitary and also in cluster and the mature ripe fruits become brown or yellow in colour and unlike the allied species, *S. incanum* in which the young green fruit at maturity change to orange-red.
- The leaves of aubergine are not eaten in this country.
- The Vavilov Institute of plant industry, Russia has maintained a rich collection of egg plant
- In India, the synthetic Bt toxin genes, cry 1Ab and cry 1b were introduced into brinjal for producing transgenics resistant to shoot and fruit borer (*Leucinodes orbonalis*). India now baned in Bt brinjal in (2010). Bangladesh still grown
- Brinjal (known as eggplant in USA and aubergine in France and UK).
- *Solanum melongena* L. (2n=2x=24) is one of the few cultivated Solanaceous species originating in the New world.
- It is known as brinjal in India.
- Acc. To Vovilov, the eggplant originated in the Indo-Burma region.
- Flower is complete, actinomorphic and hermaphrodite.
- Hetrostyly is common feature.
- Eggplant is usually self- pollinated but the extent of cross-pollination has been reported high as 29% and hence, it is classified as often cross pollinated or facultative cross pollinator.
- Out-crossing takes place with the help of insect (Bumble bee, wild bee and domestic bees.)
- Considerable heterosis for fruit yield, quality and resistant to diseases has been reported in egg plant.
- F_1 hybrids are becoming popular day by day where manual emasculation and pollination are very much in practical use.
- Historically F_1 hybrids have been common in Japan and there were about 20 hybrids in cultivation in 1935.
- Variety resistance to Phomopsis blight - Florida Market, Florida Beauty,Puha Bhairav and wild species like *S. xanthocarpum, S. indicum S. gilo, khasianum, siaymbrifolium, testiculatum.*

Brinjal little leaf

- Singh (2004) has given detailed account of this disease which is caused by phytoplasma.
- It was first reported by K.M. Thomas and C.S.Krishnaswamy in 1939 in India.
- Early symptoms are light yellow colour of leaves.

- Infected leaves are thinner. Flower buds on upright position instead of being pendulous.
- It is transmitted by leafhopper and also translated artificially by grafting.
- This disease has been seen to occur in eggplant field in kharif season in great severity in Pune.
- No resistance source is available.

Shoot and fruit borer (*Leucinodes orbonalis*)

- This insect is a major pest in India and other Asian country Caterpillar make tunnel through shoot and fruit and a large portion of harvested fruit become non-marketable.
- The wilted shoots are easily visible.

Jassid/ Leaf Hopper (*Amrasca bigutulla bigutulla*)

- Manjari Gota is reported to be tolerant.
- Infested leaf curl upwords along the margin.

Thrips (*Thrips palmi*)

- High level of resistance has been reported in wild solanum species.

Hybrid

- Pusa Bhairav (Phimosis blight resistant), Pant Rituraj, Pusa bindu
- Arka Navneet (Highest Yielding), Arka Neelkantha,
- Annamalai Aphid resistant (Annamali Agriculture University T.N.) Arka Keshav.

Others variety- Azad Kranti Arka sheel Arka Nidhi.

- Hissar Jamuni – For ratooning pupose
- Pusa Purple Long- Extra early maturing variety
- Pragati (Arka Kusumakar x Manjiri gola)
- Arka Kusumkar-Bearing finger shaped , medium long, pale green fruit in cluster
- Arka Shirish- Extra long green fruits
- **White brinjal variety** – KKM-1, Narendra Suyog (NDB white-1) and Kashi Himani

Capsicum

Scientific Name : *Capsicum spp.*

Family : Solanaceae (Nightshade family)

Origin : Mexico

2n= 2x=24.

Fruit type : Multseeded berry

- The primary centre of origin of chilli is in southern and Central America and secondary centre is Guatemala
- Among the five cultivated / domesticated species of the genus Capsicum, *C. annuum* is the most widely cultivated.
- The cultivars *C. annuum* include both hot pepper syn. Chilli (pungent fruits) and sweet pepper syn. capsicum (non pungent fruits).
- The major Chilli producing countries are India, China, and Korea etc.
- In India *C. annuum* is most commonly cultivated for commercial purpose.
- The other four species domesticated (*fruitescens, chinense, baccatum and pubescens*) are also grown in different region although very limited scale.
- The major Chilli growing state in India Andhra Pradesh, Maharashtra, Karnataka and Tamil Nadu which together consistute about 75% of total area.
- Most abundance mineral in bell peper: K
- Simultaneous emasculation and hybridization is recommended in chilli
- For maximum seed yield, ratio of seed parent to pallen parent: 3:1
- India is the largest producer, consumer and exporter of chilli among other major producers in the world and is at the top in terms of international trade.
- Sweet, nan-pungent papers are widely used at the immature are mature stage as a vegetable.
- Fruits of chilli and sweet peppers are one of the richest sources of antioxidative vitamins such as vitamin A, C and E, which protect human body from several type of cancer.
- In fact, vitamin C was first purified from papers fruits in 1928.
- Capsicum is the only genus among the plant Kingdom where the most prominent capsaicinoid, capsaicin, is synthesized in the placenta of fruits.
- Cosmetic industry uses non-pungent oleoresin (oleoresin having no capsaicin) to prepare its products.
- Oleoresin are not water -soluble, hence get accumulated in yolk and skin.
- Capsaicin extracts are used as an effective repellent against mice.

- Pepper is a dicotyledonous plant of epigeal germination.
- Fruit are non-dehiscenct, many seeded berry.
- The word 'Capsicum probably drives from the Latin word '*capsa*', meaning chest or box.
- Chilli normally considered a self pollinated crop but often crass pollination and natural cross pollination up to 90% may occur depending on bee activity and hetrostyly.
- Peppers are considered to be warm season, day neutral plants, although some line may be photoperiod sensitive
- Peepers are better adapted to worm weather then the sweet pepper, but it did not set fruit well when night temperature are greater than 24°C and below the 16°C .
- The optimum day temperature for hot paper growth range from 20° to 30 °C
- When the temperature fall below 15°C or exceed 32°C extended period, growth and yield are adversely affected.
- The optimum soil temperature for seed germination in 25°C and maximum with 30°C and the optimum temperature for productivity between 18° to 30°C.
- Capsicum (Shimla mirch) is transplanted crop approximately 150-200 g seed required for raising seedling to transplanted one ha. *Capsicum annuum* also known as sweet peeper. Dry Chilli production is also more common in southern part of the country.
- The average yield of 100-150q/ha of green chillies and 20-25 q/ha of dry chillies are obtained.
- 100 kg of fresh ripe fruits yield in about 25-40 kg dry chilli depending on the cultivar and thickness of the pericarp.
- Average yield of sweet paper varies from 150- 200 q/ha.

Variety

California wonder, Yelo wonder, Sweet Banana	**-Introduced variety**
Arka Mohini,Arka Gaurav, Arka Basant	**-Selection and Introduction from USA**

Sweet peeper hybrid Indira lorio, Bharat Puha Deepti (KT-1)

Chilli

- **Pusa Jwala-** Highly pungent tolerant to thrips and mite suitable for dry fruit production.

Pusa Sadabahar	Andhra Jyoti (G5)	Pant Shivam	Punjab Lal	Kashi Anmol
Bhagya Lakshmi (G4)	Arka Lohit	Bhaskar	Kashi Gaurav	Kashmir lang-1

Kashi Abha	Kashi Ratna- F_1	Kashi Tej- F_1	Kalyanpur Mohini	Andhra Jyoti-Cherry type fruit

Kashi sindhuri

- Bhaskar- Yellow anther type and R. to thrips and mite
- Pusa Meghdoot- First F_1 hybrid by public sector
- Kalyanpur Red-Selection from local cultivar
- Jwalamukhi- HDP and R. to virus
- Punjab Lal – For colour extraction
- Leaf curl Resistant variety-Pusa Jwala, Pusa Sadabahar, Pant-C-1

Hybrid chilli

- CH-1(MS-12 x LLS)
- CH-3 (MS-12 x 2530)
- CH-2 and Tejaswini
- Dry chilli generally contains about 6% stalks +40% pericarp +54% seeds
- Red colour of chilli is due to capsainthin ($C_{40} H_{56} O_3$).
- Pungency of chilli is due to capsaicin ($C_{18} H_{27} NO_3$) and capsicutin.
- Major pest of chilli is thrips (*Scirtothrips dorsalis*).
- NAA (Plano fix) @10ppm and Triacontanol is used to control the fruit drop.
- Anthracnose/ die back is due to fungus *Colletotrichum capsici.*
- India is the major producer, consumer and exporter of chilli in the world.
- Andhra Pradesh is the pioneer in chilli production.
- Leaf curl of chillies is due to virus and transmitted by thrips
- To prevent fruit rot spray calcium nitrate @ 1 liter/ha or broadcast of 25 kg /ha.
- Green fruit to dry fruit ratio= 10:1
- For chilli seed rate is 1-1.5 kg/ha
- Seed rate for Capsicum annuum (hybrid) is 200-250g/ha.
- Highest pungent chilli in the world : Bhoot Jolokia/Naga King Chilli (*C. chinensis*)
- Heatness of Bhoot Jolokia: 10,01,304 SHU (Scovonelle Heat Unit)
- In chilli fruit maximum pungency concentration is at placenta and septa
- Most of small highly pungent chilli: *C. fruitescens*

Potato

Scientific Name : *Solanum tuberosum* L.

Family : Solanaceae

2n=4x=48

Origin : S. America

- Potato is considered as the **king** of vegetable.
- Globally India ranks 4th in area and 3th in production.
- The Indian share of potato export is around 0.1 % of the World Export, however export is very erratic.
- It is also called Irish potato in USA.
- Operation Greens Scheme related to **TOP** (Tomato, Onion and Potato)
- The cultivation of potato on large scale was begun first in the Nilgiris Hills in 1822, thereafter; it is spread all over the country.
- Potato is grown in India in almost all the states except Kerala.
- Maximum area under potato is in Alluvial soil.
- Ridge and furrow method is mostly preferred for potato planting
- Potato flour is used for making biscuit and chapattis mixed with wheat flour.
- Potato flour can be used for preparation of idli and alu bhujia.
- Boiled, steamed or baked potatoes are more easily digested and contain more vitamin C (10-25mg/100g) then chips and french fries.
- It contains anticancerous protease inhibitors.
- The starch content is the main component of dry matter in potato.
- Suberization: Heavy of wounds dry hardening by formation of periderm
- Suberization temperature 25°C and 95% RH
- International potato centre is located at Lima (Peru), 1972

Cultivated Species

- Diploid species- (2n=2x=24) *S. steanotonum, S. phureja, S. ajanhuiri and S. goniocalyx.*
- Triploid species (2n=2x=36) *S. chaucha, S. juzepezukii.*
- Tetraploid/autotetraploid species (2n=2x=48) *S. tuberosum* subspecies *andigena.*
- *S. tuberosum* subspecies *tuberosum.* It is cultivated all over the world on commercial scale.

- Pentaploid species (2n=5x=60) ***S. curtilobun.*** It bears large violet flowers.
- Potato stems are vines are called haulms in Europe.
- The greater losses result from peeling in irregular shaped potato.
- The potato tuber is an enlarged portion of an underground branch of stem are called stolon or rhizome, **stolon is more common**.
- The tuber is formed at the tip of stolon
- These tuber and stolons is modified stem.
- The stolen tip swells when surplus food material mostly carbohydrate are diverted (stored in stolon tip) that are excess of those needed by the plants for its normal growth process.
- "Patatin" a glycoprotein, which is one of the major storage proteins in potato tuber.
- Potato cultivars have dormancy period of 4 to 8 weeks and are classified accordingly a short (below 6 weeks), medium (6-8weeks) and long (above 8 weeks) dormancy period.
- Warm temperatures during storage shorten the dormant period.
- Thiourea treatment. 1% thiourea solution for one hour.
- **GA treatment -** when Hill potatoes are to be used as seed they are cut and dipped in 1 PPM solution of GA for one hour.
- **Rindite treatment-** The mixture of ethylene chlorohydrins (7parts) ethylene dichloride (3 parts) and carbon tetrachloride (1 parts) is called Rindite.
- It is applied @ 0.3ml/kg of tubers in an airtight chamber for 4 hours.
- Potato is a cool season Vegetable crop and moderately tolerant to frost.
- The young plants grow best at a temperature of about 24°C.
- Tuberization is optimum at night temperature of 12°C.
- Optimum tuber production take place with mean temperature range of **15.6 to 18.3**°C with maximum 24°C and minimum 7°C during growing season.
- Tuber production is maximum at about 20°C temperature at about 29°C and decreased with rise in temperature.
- Temperature is about 29°C the tuber production stops totally.
- Dry matter production is most rapid at 20°C.
- Soil temperature range of 16-24°C is considered ideal for good yield.
- Soil pH less than 5.4 is not congenial for spread of common scab (*Streptomyces scabies*).
- Foliar application of urea 1% can be given to boost vegetative growth.

- Application of P and K improves the quality of tuber in terms of size and nutritional composition.
- Size of tubers for plantig varies from 25 to 40 g weight and 2.5 to 3.5 cm in diameter.
- High moisture demand occurs during tuber initiation (tuberisation) and tuber enlargement.
- The haulms are killed within 3-4 days after paraquat application.
- The aim of curing is to decrease the storage losses and to increase the shelf life of potato.
- Curing is done before harvesting as well as after harvesting.
- The skin of the tubers can be hardened by withholding irrigation at least 10 days before harvesting.
- Curing enhances periderm formation and heals harvest wounds.
- There are four grades of potato tubers *i.e.* extra large sized (above 100g), large sized (above 50g), medium sized (above 25g) and small sized (less than 25g).
- About 90% of the potatoes are produced in the Indo-Gangetic plains, and one crops is grown in a year, which is harvested in February- March.
- The potatoes for consumption and processing purpose should be stored at 8 to 12°C.
- Seed tuber are cut into small pieces having single eye each, except the pieces containing the apical bud which may have more than one eye.
- **True Potato Seed (TPS)** was developed by CPRI for multiplication of seed tubers free from viruses in North Indian Plains
- TPS concept was given by Dr. S. Ramanujam
- Dr. S. Ramanujam was the first director of CPRI, Shimla.
- The work on TPS was restarted in 1976 at CPRI and subsequently in regional office of international potato Centre (CIP), New Delhi.
- The seed rate of TPS for the raising of crop from transplanted seedling is 150g
- Seed rate of TPS for the raising of crop from seedling tuber(tuberlet) is 50g
- TPS/Botanical seed are formed in small fruit known as Berries.
- A single potato plant may have 50-100 berries.
- Single berry contain about 150-200 seeds and 1g TPS may contain 1500 to 2000 seeds.
- By using TPS, the seed cost is reduced considerably, as very small quantity of TPS (100-150g/ha) is needed as compare to seed tuber (2.5-3 tones/ha).

- The maximum production of TPS per unit area and reduced cost of seed production.
- The optimum temperature for pollen germination is about 20 degree Celsius
- Dr. Pushkarnath (1959) proposed 'seed plot technique' to grow seed in sub-tropical plains during low or no aphid conditions.
- The critical limit of ETL for a population is 20 aphid/ 100 compound leaves.
- Seed accounts for nearly 40% cost of potato cultivation.
- The aim of haulm destruction is to prevent virus infestation during high aphid population.
- Potato seed yield varies from 30-40t/ha.
- Potato tubers are harvested 10-15 days after haulms cutting, so that their skin gets hardened.
- The minimum waiting period after haulms destruction is 7 days and maximum is 21 days.
- If harvesting is delayed, the seed quality is affected due to high temperature.
- The potatoes are graded by hand or mechanical graders in 3 grade *i.e.,* large, medium and small.
- The seed potatoes are filled in gunny bags and stored in cold storages 2-4°C temperature and 90-95% relative humidity.
- Disease-free mini-tubers (Avg. wt 12g; range 5-25g)) are ideal propagules for raising healthy potato crop.
- Such mini-tubers can be produced from smaller mini-tubers (less than 5g), in vitro plantlets, micro-tubers, nodal cutting etc. under aphid- proof net-house in all potato growing zone of the country.
- In cultivated *tuberosum* and *andigena* potatoes, complete or partial male sterility is reported due to absence or poor quality pollens.
- Dry matter is usually estimated by specific gravity measurements.
- In the processing potatoes, higher solid contain or dry matter (19% or more) is preferred because product output is greater and the potato will cook faster, have a better texture and absorb less oil, if used for chip making or French fry.
- The inheritance of dry matter is polygenic in nature.

Varieties

Early – Kufri Chandramukhi, K. Lavkar, K. Ashoka, K. Jawahar, K. Surya

Mid- K. Sindhuri, K. Jyoti, K. Lalima, K. Swarna

Late –K. Deva, K. Megha

Hybrids - K. Chandramukhi (S-4485× Kufri Kuber), K.Sheetman (Phulwa× Craigs Defiance), K. Badshah (K. Jyoti×K. Alankar), K. Sindhuri (K. Red×K. Kundan)

Infestation of late blight (*phytophthora infestans*) occure both in leaves and tubers. Resistance based on R genes to specific pathotypes of the fungus has been introduced from *solanum demissum* (2n=6x=72) to the current cultivars, but the resistance can break down quickly if new race of pathotype occur.

- Potato is a self pollinated crop and propagated vegetatively. The cultivated tetraploid varieties are highly homozygous.

Introduction

- **Magnum Bonum-** From England
- **Craigs Defiance-** From Scotland
- **Up-to-Date** - Introduction from North Ireland (England) Most popular variety

Clonal Selection

- Darjeeling Red Round (DRR) Phulwa, K. Red is an excellent example of clonal selection from indigenous variety.
- **K. Safed** – Cloned
- **K. Swarna and K. Thenamalai:** Resistant to Cyst nematode and Late blight
- **K. Badshah:** Resistant to late blight, early blight and virus (MDR)
- **Kufri Chipsona I and II**: Suitable for processing purpose
- **Kufri frysona: Suitable for processing into French fries**
- **Kufri Surya:** Early maturing, heat tolerant (20°C) and hopper burn resistant
- Potato variety **Kufri Kuber** was developed from a complex cross (*solanum curtilobum × solanum tuberosum) × solanum andigena* is interspecific hybridization.
- In india, *S . demissum* has been extensively used as a source of resistance for late blight.
- The species *S. rybini and S. simplicifolium* are called **bridging species**.
- Mutation breeding has only negligible significance in potato breeding.

Physiological Disorders

Glycoalkaloids and potato greening	Exposure to sunlight
Hollow heart	Excessive N_2
Black heart	Poor ventilation
Knobbiness and secondary growth	Soil moisture deficit

Cracking	High N_2 and when the longer period of moisture deficit is followed by irrigation or rain.
Enlarged lenticels	O_2 deficiency/excessive soil moisture
Black spot	Low in K, high in phenol compounds and tyrosine
Internal brown spot	Moisture deficiency
Chilling injury	0°C
Freezing injury	-1 to -2°C
Sprouting in storage	Major problem of storage
Poor or uneven sprouting in field	Planted just after the removal from cold storage

Diseases and Pests of Potato

Diseases	**C.O.**	**Remarks**
Fungal		
Late blight	*Phytophthora infestans* (Replant pathogen)	Most serious and epiphytic
Early blight	*Alternaria solani*	Concentric spot / Targeted board
Potato wart	*Synchytricum endobioticum*	Tuber borne, Endemic in Darjeeling hills. Resistant var. Kufri Kanchan
Black scurf	*Rhizoctonia solani(Re-emerging) /Pellicularia filamentosa*	-
Dry rot	*Fusarium spp.*	Storage disease
Bacterial		
Soft rot	*Erwinia carotovora*	-
Potato scab	*Streptomyces spp.(G^+bacteria)*	Alkaline soil - Favourable
Bacterial wilt	*Ralstonia solanacerum*	-
Viral		
Potato spindle tuber	***Viroid*** *(without protein coat virus, consist of circular necked RNA)*	Discovered by **T.O.Diener (1971)**
Potato leaf roll	*Virus*	Aphids
Latent or faint mosaic	*PV-X and S*	Aphids
Severe mosaic	*Virus*	Aphids
Nematode		
Cyst / Golden Nematode	*Globodera rostochinensis*	Serious problem in Niligiri hills. Resistant var. Kufri Swarna, K. Dewa
Pests		
Potato tuber moth	*Pthorimea opercullela*	30-70% damage in tuber during storage
Aphids	*Vector of potato virus*	
Cut worm	*Agrotis ipsilon*	
Mite	*Hemitarsonemus latus*	Major pest in Deccamn Plateau

CHAPTER-4

Okra

Scientific Name : *Abelmoschus esculentus*

Family : Malvaceae

Origin : Tropical Africa

2n=130

- Okra is warm season crop.
- It is a tropical crop.
- Okra has caducous calyx (**Shedding type**).
- India ranks first in world in okra production.
- It is sensitive to frost.
- Major producing state in West Bengal followed by Bihar.
- High iodine content of fruits helps to control the goiter while leaves are used in curing inflammation and dysentery.
- Protogyne condition found in okra.
- The roots and stems are used for clearing cane juice in preparation of "Gur" and "Jaggary"
- Okra is also called as:
  - ☆ Lady's finger in England
  - ☆ Gumbo in USA
  - ☆ Quimgombo in Spanish
  - ☆ Bhindi in India
- Mucilase present in okra fruits is polysaccharides *i.e.* galacturonic and glucuronic acids.
- Okra has the highest chromosome number among the vegetables that is 2n=130.

- The most commonly reported chromosome number of okra is 2n=120-130.
- Cultivated okra (*Abelmoschus esculentus*) is an artificial amphidiploids of (*A. ficulneus × A. tuberculatus)*
- An okra cultivar from Ghana, *A. manihot* ssp. manihot has the highest chromosome number 2n=194.
- *A.manihot* is cultivated mainly for consumption as a leafy vegetable.
- *A. moschatus* is better known as **musk mellow**, as the musk-scented perfume made from its seeds.
- The mucous leaves of this species are also used as leafy vegetable.
- Okra is natural amphidiploids or allopolyploid.
- Okra is often cross pollinated plant (extent of cross pollination is 4-19%).
- For *A.esculentus* 0-69% out-crossing has been reported in Puerto Rico.
- Ovary is superior.
- Fruit of okra: Capsule
- Resistance to YVMV has been the major breeding objective in okra.
- Dr. Harbhajan Singh initiated systematic research work on improvement of okra in India.
- Genus *Abelmoschus* was first established by Medikus in 1787.
- Okra has an indeterminate growth habit.
- Hybrid seed of okra on commercial scale is mostly produced in Ranebennur area of Karnataka and Buldha district of Maharastra.
- Okra is commonly used for tender pods.
- Frozen okra is also exported to U.K.
- Okra, *Abelmosschus esculantus* was earlier botanically Named as *Hibiscus esculentus.*
- In genus *Abelmoschus,* calyx, corolla and stamens are fused together at the base and fall in single piece after anthesis (deciduous nature), where as in case of *Hibiscus* these floral parts are persistence.
- The genus *Abelmoschus* is small and consist of at least six species (Borssum Waalkes,1966)
- Acc. to new classification, the *Abelmoschus* comprises nine species.
- *Tubercular esculantus* is combine characteristics of *A. tuberculatus* and *A. esculentus 2n=182)*
- Flowers are hermaphrodite and actinomorphic.
- First flower may appear at 5th to 7th nodes.

- The greatest increase in fruit weight, length and diameter occurs during 4th to 6th days after pollination.
- Pigmentation of calyx, corolla and fruit colour is under monogenic control.
- Yellow vein mosaic virus (YVMV) disease is control by two complementary dominant genes Yv_1 and $Yv_{2.}$
- Pungens carries true resistance to YVMV disease and cold fruit set species.
- Fiber formation in the fruit starts from the 5th to 6th days after fruit formation.
- Commercial varieties have five ridges.
- Fruit consists of 55-62% pericarp, 30-40% seeds and 7-11% axil.
- **Seed rate:**
  - ☆ **Summer season crop:** 18-20kg/ha.
  - ☆ **Rainy season crop:** 10-12kg/ha/
- **Seed viability:** 2 years.
- Optiumum temperature for growth, flowering and fruiting is 24-27^{0}C.
- Seed does not germinate below 17^{0}C.
- Temperature higher than 42^{0}C may cause flower drop.
- Okra pods become ready for first harvesting after 45 days of sowing.
- Best time of picking being 6 to 7 days after anthesis.
- Isolation distance for Breeder/Foundation seed 400m and Certified seed 200m.

Seed yield:

- ☆ 1000 kg/ha.
- ☆ 1000 seed weight-50g.

- Export standard – 6-8cm long fruits.
- Pusa Sawani and Parbhani Kranti – Suitable for export.
- Ratoon cropping is done in okra.
- Varieties suitable for rationing- Arka Abhay and Pusa A-4.
- Perkins Long Green- recommended for hilly areas.
- Source of resistance to root-knot nematode has been reported in cultivar Clemson Spineless.

Crop Improvement

1. Introduction

- Clemson Spineless
- Louisiana,
- Velvet Round
- Perkins Long Green- recommended for hilly areas.
  - ☆ A cultivar from Ghana (Africa) known as *A. manihot ssp. manihot* now known as *A. caillei* a semi-wild species introduced into India has been successfully used as a sourse of resistance to YVMV in breeding of Parbhani Kranti and Punjab Padmini.

2. Pureline Selection

- Pusa Makhmali
- ujarat Bhindi No-1

3. Hybridization

- Pusa Sawani (Pusa Makhmali × IC-1542), day neutral variety, Spineless
- Punjab Padmini (*A. esculantus* × *A. manihot ssp. manihot*) backcrosses were made.
- Parbhani Kranti (*A. esculantus* × *A. manihot ssp. manihot*)
- Arka Anamica (*A. esculantus* × *A. manihot ssp.tetraphyllus*)
- rka Abhay (*A. esculantus* × *A. manihot ssp. manihot*)
- Varsha Uphar (Lam Sel. -1 × Parbhani Kranti)- resistant to YVMV.

Others- Punjab-7, Hissar Unnat, Selection-2-3, Shitla Uphar, Shitla Jyoti, Kashi Bhairav, Kashi Mahima.

4. Mutation

- MDU-1 –Induced mutant from Pusa Sawani
- Punjab-8 (EMS-8) - Induced mutant from Pusa Sawani by treating seed with EMS 1%, resistant to YVMV and tolerant to fruit borer.

Others varieties-Pusa A-4, CO-1, Kashi Pragati, Kashi Mohini, Kashi Mangali, KashiVibhuti, Kashi Satdhari, Kashi Lila.

- Kashi Lalima, Red Bhindi- Both is red colour variety.

Pest and Disease

Major pest of okra is: Fruit borer (*Earias vitella*).

Blister beetle: (*Mylabris pustulat*a) is a pest of okra

Disease

Yellow Vein Mosaic (Geminivirus)

- ☆ Most devastating disease of okra.
- ☆ First reported by C.S. Kulkarni in 1924.
- ☆ Characteristic symptoms include homogenous interwoven network of yellow veins enclosing islands of green tissues.
- ☆ Initially, the infected leaves exhibit only yellow colour veins but in later stage, entire leaf turns completely yellow.
- ☆ Trasmission is by Whitefly *(Bemisia tabaci)*
- ☆ White fly population and YVMV incidence are maximum during March-June under Bangalore and Pune condition.

- ❒ In north India, severity is high in rainy season planting, particularly in June sowing.
- ❒ Arka Anamika, Arka Abhay and Parbhani Kranti were initially highly tolerant.
- ❒ *manihot* accessions have shown resistance to YVMV.

Enation Leaf Curl

- ❒ First observed at IIHR, Bangalore in 1924 by S.J. Singh.
- ❒ There is twisting of main stem, lateral branches and leaf petiole.
- ❒ Virus transmission is by grafting. Natural transmisstion is by whitefly.
- ❒ No resistant source is reported.

Cercospora Leaf Spor (*Cercospora abelmoschi*)

- ❒ The disease occurs in humid season.

CHAPTER-5

Cole Crops

General

- The word 'Cole' seems to have come from the word "*Caulis*" meaning stem.
- Cole group plants were oringinated from a single wild *Brassica oleracea var. (sylvestris* L.), commonly known wild cabbage.
- Cole crops are rich source of Vit-A.

Cole crops	Vit-A (per 100g)
Cabbage	400 IU
Cauliflower	70 IU
Knolkhol	20 IU
Brussels sprouts	640
Sprouting Broccoli	9000IU
Kale	20000IU

Brassica cretica is the probable ancestor of the cauliflower.

- Brussels sprouts were probably developed near Brussels in Belgium during the 14th century.
- Thousand head kale-*B. oleracea var. ramosa.*
- Cole crops need chilling requirement for emergence of inflorescence

Crops	Chilling Temp.	Weeks
Cabbage	4-10 °C	6-8
Brussels sprouts	4-10 °C	4-6
Knolkhol	7-10 °C	5-7

Cole Crops are Grown in Calcicole Soil

- Among the cole crops in india having maximum area is cauliflower
- Cole crops are highly heterozygous and heterogonous in nature
- Cole crops are highly cross pollinated crops due to protogyny and self incompatibility

- Pollinating agent – Bees and flies
- Tetradynamous is special anther type feature
- Kitchen kale – *B. oleracea var. fimbriata.*
- White cabbage – *B. oleracea var capitata* L. f. *alba.*
- Savoy cabbage - *B oleracea var. sabauda* (Chieftem)
- Red cabbage – *B. oleracea var capitata* L. f. *rubra.*
- Curly kale – *B. oleracea var. acephala* sub var. *laciniata*
- The book Cole Crops was written by Nieuwhof (1969).
- All Brassica spp. contains glucosinolates which in crushed leaves are broken down by the enzyme myrosinase to give bitter taste and goitrogenic substance.
- The family cruciferae is characterised by 4 petals.
- Fruit is special kind of pod called siliqua.
- 'U' - triangle was given by Nagaharu in 1935.
- Interspecific hybridization between cabbage and Chinese cabbage has been done to breed soft rot resistant character in Chinese cabbage.
- New vegetable **Hakuran** (Cabbage × Chinese cabbage) from cruciferae.
  - ☆ By embryo culture
  - ☆ Hakuran is highly resistant to soft rot disease and highly resistant to drought and heat.
- **Raphanobrassica** : (Radish × cabbage)
- **Swede** : (Turnip × Cabbage) cruciferous root vegetable

CABBAGE

Scientific Name : *Brassica oleracea var. capitata*

Family : Cruciferae

Origin : Mediterrnean region

2n =18

- Flavour of cabbage leave is due to glucosides "**Singrin**"
- Cabbage is shallow rooted crop
- Edible part of cabbage: Head
- Type of inflorence is katkin
- Head compactness is determined by pearson formula **$Z=(C \times 100)/W^3$**
- Anticancer property is due to the presence of Indole -3- carbinol

- Cabbage cultivars can be classified into three broad groups:

(1) White cabbage	*B.O. var. capitata f. alba*
(2) Red cabbage	*B.O. var. capitata f. rubra*
(3) Savoy cabbage	*B.O. var. capitata f. sabauda*

White1 cabbage is of commercial importance in India.

- Early cultivars like- Golden Acre and Pride of India are both selection from Copenhagen Market which was 1st introduced in Denmark.
- **Pusa Ageti**- 1st tropical variety developed for cultivation in high temperature conditions.
- **Qiangan-1** is a new cultivar developed in China for spring season.
- **September** is a cultivar of cabbage.
- **Pusa Drum Head** cultivar is tolerant to 'black leg' disease.
- Pointed head white cabbage cultivars – Charleston
- Cultivar **Majabi** and **Pusa Mukta** are most resistant to black rot disease which is caused by *Xanthomonas campestris.*
- Red cabbage cultivars - Red Acre
- Savoy cabbage cultivars - **Perfection** (syn. Chieftain),
- Hybrid cultivars of IAHS (Indo-American Hybrid Seeds), Bangalore Ganga, Kaveri, Yamuna.
- **Yanchun** is an early spring cabbage cultivar.
- **Karino** (hybrid cultivar) is resistant to bolting.
- Cultivar Midore highly tolerant of black rot disease.
- **Pusa Sambandh**- suitable for high density planting (HDP).
- A measure of salt content is called C-index) which indicates the number of gram of salt / lit of soil moisture.
- C-index according to crops -

White cabbage and Savoy cabbage	C_6
Red cabbage	C_4
Kale	$C_2 - C_4$

Pusa Drum Head (flat type) is the most tolerant cultivar to saline

- Black leg disease is more common in saline soil
- Speedling sowing of seeds in thermocole plates followed in cabbage.
- Cry1A gene resistance against cabbage butterfly and Diamond back moth (DBM)

- Spray of CCC/SADH (2500-5000ppm) increase the low temperature resistant in cabbage
- Sauerkraut is fermented product of shredded cabbage
- Seed production in cabbage is done by Head to seed method
- Intercropping of tomato with cabbage can reduce the damage of DBM

Seed rate – Early 500-600g

Late 374-400g

- **Diamond back moth (DBM)** : (*Plutella xylostella*)
  - ☆ It is serious pest of cabbage
  - ☆ Larvae feed on leaves and later make galleries in head
  - ☆ Indian musterd which attract 80-90% DBM used in trap crop

CAULIFLOWER

Scientific Name : *Brassica oleracea var. botrytis*

Family : Cruciferae

Origin : Cyprus (West Asia)

Near Mediterranean region

2n =18

- Thermo-sensitive crop
- Edible part : Curd (prefloral fleshy apical meristem)
- Cauliflower was introduced in India during 1822.
- Scooping is the removal of central portion of curd for easier initiation of flower stalk in cauliflower.
- Tropical cultivars form curd at temperature 20°C or above.
- Temperate cultivars form curd at temperature 10°-16°C
- A premature initiation of flower buds is characterised by riceyness in cauliflower
- Reliance is a cultivar of cauliflower.

Type of crop according to maturity –

Early		Mid- early	Mid -late	Late
Curd initiation temperature				
Early I (20-27°C)	**Early II (20-25°C)**	**16-20°C)**	**12-16°C**	**10-16°C**
Pusa early synthetic	Pusa Deepali	Pusa Hybrid -2	Pusa Synthetic	Pusa Snowball-1,2,16
Pusa Karthik Sankar (F1)	Pusa Ketki	Pusa Sharad	Pusa Himjyoti	Pusa Snowball-K1

Improved Janpanese is a cultivar introduced from Israel.

- Pusa Snowball – K-1 -tolerant to black rot.
- Self blanched variety- Pusa Deepali
- Synthetic variety- Pant Gobhi-3

Physiological disorder

Disorders	Symptoms	Reasons
1. Ricyness	Premature initiation of floral buds on upper surface of curd	Fluctuation in temperature, High Humidity, Excess Nitrogen
2. Fuzziness	Velvety appearance of pedicel	Cultivation in abnormal time.
3. Blindness	Without terminal bud	Frost
4. Leafyness	Small green leaf in the curd	High temperature
5. Buttoning	Development of small curd in the young plant	Nitrogen deficiency, sowing early variety in late season
6. Hollow stem	-	Excessive Nitrogen
7. Whiptail	Mid rib development	Molybdenum deficiency, common in acid soil
8. Chlorosis	-	Magnesium deficiency
9. Brown rot/red rot	Formation of cavity and hollowness of the stem	Boron deficiency
10. Brown heart	-	Excess CO_2

SPROUTING BROCCOLI

Scientific Name : *Brassica oleracea var. italica*

Family : Cruciferae

Origin : Mediterranean region

2n =18

- The Name 'broccoli' refers to the young shoots, *brocco* is an Italian word which mean shoot.
- Broccoli rape - the sprouts which develop on turnips.
- Rich source of sulphoraphane- compound associated with reducing risk of cancer.

- It has 130 times more vitamin A than cauliflower and 22 times more than cabbage.

Cultivars

Palam Kanchan	Yellowish green , heading broccoli
Palam Vichiyra	Purple, heading broccoli
Bronzino	Purple, heading broccoli
Palam Haritika	Green, Sprouting broccoli
Calabrese	Green, Sprouting broccoli
PalamSamridhi	Green , Sprouting broccoli, Early variety
Pusa KTS-1	Sprouting broccoli
Solo Head	Self blanched
De Cicco	winter
Palam Uphar	
Mercedes	

Solohead variety resistant to *Xanthomonas campestris.*

- **Green Valiant -** stable and better adepted to unfavourable environments.
- The head may be 15-25 cm in diameter and weight is about 250-600g.
- Yellwing of broccoli in storage is mediated by ethylene action and 1-methyl cyclopropene treatment inhibits yellowing and reduces respiration.

BRUSSELS SPROUT (Mini Cabbage)

Botanical Name : *Brassica oleracea var. gemmifera*

Family : Cruciferae

Origin : Mediterranean region

2n =18

- Cold and moisture loving , frost resistant crop
- Edible part swollen axillary bud (sprouts/buttons/mini cabbage)

Varieties

Hilds ideal	Rubine (F_1 hybrid)	Jade cross	Amager Market	Long Island
Early Mom	Frontier Zuerg	Danish Prize	Catskill Dwarf	Dwarf improved

Excessive Application of Potash imparts bitterness, affects quality

- Topping is done to increase the harvesting time
- Loose sprout marketed as a "blowets"

KHOLRAB/KNOL KHOL

Scientific Name : *Brassica oleracia var. gongylodes,*
2n =2x=18,

Family : Cruciferae

Origin : Mediterranean region

- Kholrabi is the German Name for Cabbage- Turnip
- Edible part of knol-khol is swollen stem called *"tuber or knob"*
- Stem tuber are developed at ground level
- In India knol-khol is more popular in **Kashmir**
- In Kashmir knol-khol leaves are also used as green
- Best time for planting – October
- Seed rate – 800-1000g/ha
- Early varieties of knol-khol are more prone to premature bolting (Purple varieties)
- Generally, tubers of 5-8 cm diameter favoured in the market
- Individual tuber may be weight upto 200-250g

Important varieties

White vienna	Purple vienna	King of North	Golithwhite (Sadashiv)	Large green

New variety – Palam Tenderknon and **Pusa Virat**

- Purple speck- earliest variety
- In kale, curly type is used for vegetable purpose.
- Kale is more salt tolerant than cabbage.
- Chinese kale *Brassica alboglabra*
- The flowers are white.
- Chinese kale is a typical tropical or hot weather vegetable.

CHINESE CABBAGE

Scientific Name : *Brassica oleracia var. chinensis*

Family : Cruciferae

Origin : China,
2n =2x=20

- Non-heading Chinese cabbage *B. campestris sub sp. chinensis.*

- Heading Chinese cabbage – *B. campestris sub spp. pekinensis.*
- Pak-choi- non-heading Chinese cabbage.
- **Asveg-I** F_1-hybrid from AVRDC can be grown in tropical and sub-tropical region.
- In Pak-choi, occurance of broad white petiole is common which is a characteristic feature only of Chinese cabbage (ssp. *pekinensis*).
- A new F_1-hybrid Zhengbai No. 4 has multiple disease resistance.
- Tipburn is one of the most common physiological disorders of Chinese cabbage.
- 'Tipburn' - due to Ca-deficiency.

KALE

Scientific Name : *Brassica oleracia var. acephala*

Family : Cruciferae

Origin : Mediterranean region,

2n =2x=18

- The flowers are white
- Chinese kale is a typical tropical vegetable
- Kale is a more salt tolerant than cabbage

CHAPTER-6

Cucurbits

General

- Largest number of cultivated vegetables belonged to family: Cucurbitaceae
- Cucurbits are the largest group of summer vegetable crops.
- Cucurbits- term coined by Dr. Bailey
- Fruits is pepo (inferior)
- The fruit peduncle is characteristic feature in cucurbita spp.
- Cucurbits are generally richer in methionine than the legumes
- Seed of cucurbits are exendospermic
- *Cucumis* and *Cucurbita* have usually ordinary unsaturated acids like oleic and linoleic acids
- Vegetative propagated Cucurbits: Parwal (Pointed Gourd), Chow-Chow, Kundsru (Ivy Gourd)
- Ivy Gourd is propagated by **root cutting**
- Kakrol is rich in protein and Chow-Chow fairly high in Ca.
- Cucurbits used for orNamental purpose is: Vegetable marrow
- Bitter gourd rich in Vitamin-C (96 mg/100 g), Pumpkin containing high carotenoid, Kakarol high in protein (3.1%), Chow-Chow fairly high in Calcium (140 mg/100g)
- **Metaxenia** common in cucumber and bottle gourd
- Bitter pollen or foreign pollen fertilizes the non-bitter or normal ovule causes bitter fruit known as metaxenia
- Bitter principle in cucurbits due to the presence of 'cucurbitacins' *i.e. tetracyclic triterpenes*
- Cucurbitacins and terpene compounds are responsible for bitter taste and flavour

- Watermelon is a highly cross pollinated crop due to separation of sex-monoecious and andromonoecious
- Flowering in cucurbits generally starts 40-45 days after sowing (DAS)
- Sex ratio in cucurbits: 25-30:1 or 15:1
- Hermaphrodite is considered as primitive sex form in cucurbits
- Long day and high temperature promotes male flower in cucurbits
- 1st 4-6 nodes bear staminate flower than pistillate will appears in cucurbits
- Staminate flower appear 7 days earlier than pistillate in cucurbits
- Growth regulator application in cucurbits is done at 2-4 leaf stage.
- Because activity of florigens and their response will be more at this stage
- Generally in cucurbits to enhance the female flower production: Ethylene is used
- Short day treatment 9 hrs: Female flower is more
- Generally in cucurbits to enhance the male flower production: GA_3 is used
- Long day treatment 16 hrs: Male flower is more
- Pollination is entomophilous

CUCUMBER

Scientific Name : *Cucumis sativus*
2n-2X=14,

Family : Cucurbitaceae

Origin : India

- Day neutral plant, Thermophilic crop
- 1st sequenced vegetable crop
- 2nd most widely cultivated cucurbit after watermelon
- White heart at central position shows poor quality fruits
- Prefers slightly low temperature than watermelon and muskmelon
- Tolerates cooler weather than melons
- It is cultivated either for fresh consumption as slicing cucumber, or as pickling cucumber for preservation, marinated with vinegar, salt, dill or other spices
- Cucumber has narrow genetic base (3-8% polymorphism).
- Most common sex form: Monoecious
- Palam Khira with small-sized is pikling type grown in Saharanpur (UP)

- A local cultivar Bharkin endemic of Sikkim
- Harvesting -full mature stage
- Most favourable temperature for cucumber: 18-24°C
- Temperatures required for the seed germination of cucumber ranges between 15-35°C
- In cucumber bitterness in fruit is due to "cucurbitacins"
- Chemically cucurbitacin is tetracyclic triterpenes
- Progenitor of cultivated cucumber: *Cucumis hardwickii*: 2n=14
- African horned cucumber *Cucumis metuliferus*
- West Indian Gherkin: *Cucumis anguria*
- Hedge hog or Teasel Gourd *Cucumis dispaceus*
- ***Cucumis hystrix***: Resistant to downy mildew, gummy stem blight, virus and nematode
- **Source of B-carotene (Orange fleshed) species:** *Cucumis sativus* L. var. *xishuangbannanesis* (Ore)
- **New synthetic species:** *Cucumis x hytivus* 2n=2x=38, synthetic allotetraploid
  - ☆ Derived from interspecific hybridization between (*Cucumis hystrix x Cucumis sativus*) through embryo culture and followed by chromosome doubling
  - ☆ **Genome:** HHCC (Amphidiploid)

Fresh market/slicing/salad cucumber

- **Fruit shape-** cylindrical with blocky or rounded ends
- **Fruit length/diamatere ratio:** more than 4 (larger LD)
- **Fruit skin colour:** dark-green colour

Parthenocarpy (Seedless cucumber)

- Parthenocarpy is the ability to develop fruits without pollination
- Complex inheritance or incomplete dominant gene (P)

Pickling cucumber

- **Fruit shape:** cylindrical or tapered shape
- **Fruit length/diamatere ratio:** 2.8 to 3.8 (smaller LD)
- **Fermented organism in brine solution:** *Lactobacillus sp.*

Gynoecious cucumber

- ☆ Produces only female flowers Male flowers artificially induced through $AgNo_3$ male and morphologically bisexual flowers in gynoecious lines Gynoecious in cucumber is controlled by single dominant gene (F/Acr)
- ☆ 1st gynoecious F_1 hybrid, Pusa Sanyog, developed in India in 1971

- ❒ Economic sex ratio of cucumber: 15:1
- ❒ The expression of sex forms is controlled by a series of multiple alleles at the "F/M" locus interacting with genes controlling the photoperiodic reaction
- ❒ Induction of parthenocarpy in cucumber: **Choloroflurenol**
- ❒ Ethylene sex hormone affect the sex differentiation in cucumber
- ❒ **AVG** induce only male flowers in cucumber
- ❒ **Seed rate:** 2.5-4 kg/ha
- ❒ Cucumber fruit yield reduction is due to crown fruit inhibition or 1st fruit inhibition
- ❒ Xenia and metaxenia commonly occurs in cucumber
- ❒ **Xenia:** Effect of genes from the male parent on the development of fruit or seeds
- ❒ **Metaxenia:** Effect of pollen on fruit shape and other fruit characteristics
- ❒ External fruit quality characteristics governed by 1-3 genes
- ❒ The most effective method for the improvement of quantitative traits, such as yield in cucumber, may be recurrent selection.
- ❒ Single-seed-descent method is a modification of pedigree breeding
- ❒ Development of inbred lines from an F_2 population in cucumber mostly breeders uses pedigree method
- ❒ More stable female sex expression in F_1 hybrids: gynoecious x gynoecious and gynoecious x hermaphroditic

Pest and Diseases

Varieties	Breeding methods	Specific features
Japanese Long Green	Introduction	Introduction from Japan -
Straight-8	Introduction	Introduction from USA -
Pusa Uday	Selection from local cultivar	Suitable for throughout the year
Pusa Barkha	Extra early variety	Tolerant to high temperature, downy mildew
Swarna Ageti	Hybridization	Slicing type

Swarna Sheetal	Hybridization	Slicing type
DCH-1 and DCH-2	Hybridization	Tropical gynoecious hybrids
Himangi	Poinsette x Kalyanpur Ageti	MPKV, Rahuri
Phule Shubangi	Poinsette x Kalyanpur Ageti	MPKV, Rahuri
Sheetal	-	-
Pant Khira-1	Selection	Resistant to PM
Pant Parthenocarpic Khira-2		Parthenocarpic variety
Pant Parthenocarpic Khira-3		Parthenocarpic variety
F_1 Hybrid:		
Pusa Sanyog	Japanese gynoecious line x Green Long Naples	Only one F_1 hybrid release by IARI

Varieties of Cucumber

- Downy mildew resistant cucumber varieties: Palmetto, PR-27, Santee and Palomar
- Cucumber mosaic resistant varieties: Shamrok Ilima, Ohio MR-200, Ohio-MR-17 and Wisconsin SMR-9
- Adromonoecious variety: White Lemon (Australian variety)
- Fruit fly is very serious pest of most of the cucurbitaceous vegetables
- Cucumber mosaic is transmitted by aphids
- Bacterial wilt of cucumber is caused by bacterium (*Erwinina tracheiphila*) transmitted by striped cucumber beetle
- Gummy Stem Blight: *Didymella bryoniae*, resistance source- *C. hystrix*
- Powdery mildew of cucumber: *Sphaerotheca fuligenea*
- Angular leaf spot (bacterial disease caused by*Pseudomonas lachrymans*)
- Pillow is the physiological disorder of cucumber due to calcium deficiency
- Chilling injury as a physiological storage disorder of cucumber when exposed to below 15°C

MUSKMELON (Wholesome Food/ Kharbooz)

Scientific Name : *Cucumis melo*

2n=2x=24,

Family : Cucurbitaceae

Origin : Tropical Africa (Sahara Desert)

- It have a good inter-state trade in India

- Melon used as vegetable: Round melon and Long melon.
- Immature melons may be used fresh in salads, cooked or pickled
- Mature fruit may be eaten fresh as a dessert fruit
- Edible portion of melons contain water 90% and CHO 10%
- Cantaloupe contains 45 mg and Honey Dew 32 mg of Vitamin-C per 100 g of edible portion
- Muskmelon mixed crop with sugarcane in Uttar Pradesh
- Variety Sarda melon is grown in Afghanistan and is available in India in October-November
- Muskmelon seed doesn't germinate at temperature lower than 18°C
- Muskmelon is slightly more tolerant to soil acidity
- Predominant sex form in muskmelon: Andromonoecious
- Hand pollination is necessary for andromonoecious types
- Ideal for sugar accumulation: cool nights and warm weather
- High quality melons should have TSS: 12-15% or more

Seed rate 5-6 kg/ha

- Melon is a diploid species
- Fruit set in monoecious lines of *Cucumis melo* is 29-42%
- Male sterile line (ms-5) has been used for production of commercial exploited for F_1 hybrid production
- Ideal accumulation of sugar in the muskmelon fruits: Cool night and warm days
- The yellow and orange-fleshed melons contain more than 350 mg of B-carotene, a precursor of vitamin A
- 1st Horticultural classification of melons was given by Naudin (1859) and modified by Munger and Robinson (1991)
- *melo* is the most variable species of the genus Cucumis
- **Snap melon**: *Cucumis melo var. momordica*: Resistant to DM, PM and CGMMV
- **Snap melon variety:** Pusa Shandar
- Snap melon and musk melon are intercrossable
- **Kakri/Vellaraikkai:** *Cucumis melo* var. *utilissimus:* Used as salad.
- Snakemelon (*var flexuosus*) is important salad-type melon in north India.
- **Aromatic melon:** *Cucumis melo* var. *reticulatus*

- **Resistance to fruit fly** (*Dacus cucurbitae*): *Cucumis trigonus* (syn. *C. callorus*)
- **Resistant to nematode:** *Cucumis anguria, Cucumis ficifolia and Cucumis metullferus* are
- **Gooseberry gourd:** *Cucumis myriocarbus*
- Harvested at full slip stage
- Long distance marked harvested at half slip stage
- Full slip stage of muskmelon contains maximum sugar
- Climacteric fruit

Varieties of Muskmelon

Varieties/Hybrid	Breeding methods	Remarks
Arka Jeet	Selection	High TSS variety
Arka Rajhans	Selection	Tolerant to powdery mildew
Pusa Madhuras	Selection	-
Pusa Sharbati	Kutana x PMR-6 (USA)	-
Pusa Sarda	-	First variety of Sarda melon, suitable for net-house under north Indian plains
Kashi Madhu	-	Long storage capacity
Punjab Raseela	C. melo var. momordica (Phoot) x Indian cultivar	Moderately resistant to downy mildew
Punjab Sunehri	Hara Madhu x Edisto	-
Hisar Madhur	Selection	-
Hisar Saras	Selection	-
Hara Madhu	Selection	Netting stage
Durgapura Madhu	Selection	-
MHY-5	Durgapura Madhu x Madhu Hara	-
Jobner 96-2	Selection	-
F1 Hybrid:	-	-
Pusa Rasraj	M-3 x Durgapura Madhu	Utilization of monoecious line
Punjab Hybrid	MS-1 x Hara Madhu	1st F_1 hybrid
MH-10	Gynoecious line cultivar Indian	

Specific features of important variety:

- **Non slip stage variety:** Hara Madhu
- **New long melon variety:** Pusa Utkarsh
- **Resistant to Fusarium wilt:** Golden Gopher, Iriquois, Minnesota Midget, Harvest Queen, Delicious 51
- **Exotic varieties:** PMR-45, Jacumba and Campo

Important pest and disease

- **Powdery mildew:** The most prevalent species is ***Podosphaera xanthii***
- **First variety resistant to powdery mildew:** PMR 45 followed by PMR-6, 7
- **Downy mildew:** *Pseudoperonospara cubensis.*
- Major problem in temperate and tropical regions with high relative humidity.
- **Fusarium wilt:** *Fusarium oxysporum* f. sp. *melonis* is a soil-borne disease causing (Four races have been reported.
- **Two resistant genes:** Fom-1, Fom-2)
- **Sudden wilt disease of melon:** *Monosporascus cannonballus*
- **Gummy stem blight:** *Didymella bryoniae*-major problem in hot and humid conditions (tropical, subtropical regions and greenhouses).
- CMV, CGMMV, SQMV, PRSV, and ZYMV are prominent in the spring (dry season) crop of cucurbits

Melons	
Common Name	**Botanical Name**
Musk Melon	*Cucumis melo*
Long/Serpent melon/Kakri	*Cucumis melo var. utilissimus*
Oriental pickling melon	*Cucumis melo var. conomon*
Snap Melon/Phoot/Phoonte Melon	*Cucumis melo var. momordica*
Nutmeg melon/Netted melon/Persian Melon	*Cucumis melo var, reticulatus*
Serpent melon/Snake melon	*Cucumis melo var. flexuosus*
Mango melon/Garden melon	*Cucumis melo var. chicko*

WATER MELON

Botanical Name : *Citrullus lanatus*

Family : Cucurbitaceae

Origin : Tropical Africa

2n=2x=22

- Fruits contain 95% water.
- Used as a dessert and its juice is refreshing in summer
- White heart at central position shows poor quality fruits.
- Harvesting at full mature stage.
- Antheracnose + Powdery mildew + downev mildew resistant variety- Arka Manik

- Cucurbitacin E is main bitter susbstance (*Tetracyclic teiterpenes*).
- Potasium contains 160 mg per 100g
- Arka Manik is suitable for long distance transport.
- *C. naudinianus* cucurbitacin –B and E are present
- Watermelon when grafted on bottlegourd cv. Renshi the incidence of Fusarium wilt is uncommon

Varieties

(A) Introduction

- Asahi Yamato – Japan
- Sugar Baby- introduction fromUSA (11-13%TSS)
- Improved Shipper- introduction fromUSA
- New Hampshire Midget- introduction fromUSA

(B) Selection

- Durgapura Meetha
- Durgapura Kesar- Yellow fleshed variety
- Pusa Rasal
- PKM-1

(C) Hybrid

- ☆ **Arka Manik:** IIHR-21 x Crimson Sweet(USA)
- ☆ **Arka Jyoti:** IIHR-20 x Crimson Sweet (USA)
- ☆ **Pusa Bedana (triploid):** Tetra2 (4x) x Pusa Rasal (2x)- Seedless Variety
- ☆ It is released by Dr. Kihara. 1972
- ☆ **Durgapura Lal**: Sugar Baby x K-3566

- Yellow Gold-48- Yellow fleshed variety Developed by Bayer Company
- Seed maturity and edible maturity coincide with each other.
- Yield losses upto 30% have been observed due to weeds.
- Metallic sound at the time of Harvesting shows: immaturity
- Heavy dull sound at the time of Harvesting shows: maturity

BOTTLE GOURD

Botanical Name : *Lagenaria siceraria*

Family : Cucurbitaceae

Origin : South Africa

2n=2x=22

- Basic chromosome number, x=5 and n=5+5+1 (2n=22)
- Kofta and Petha are most popular preparations from bottle gourd
- Highly sensetive to photoperiod.
- Dry shell is used for preparation of musical instrument.
- Processed product tooty-fruity
- Fruit pulp is very good source of fibre free carbohydrates.
- **Germination:** 25-30°C Brownish or whitish brown seeds germinate well
- High temperature and High N_2 induce maleness in Bottle gourd
- Heterosis is present through its life cycle in all stages.
- *Perennial* spp.- *Lagenaria sphaerica*

Varieties

(A) Selection:

- Pusa Summer Prolific Long
- Pusa Summer Prolific Round
- Arka Bahar
- Pusa Naveen
- Pusa Sandesh
- NDBG-1
- Arka Bahar
- Kalyanpur Long Green
- Narendra Madhuri
- **Samrat** - Highest Yield

(B) Hybrid

- **Pusa Meghdoot:** (Pusa Summer Prolific Long x Sel.-2)-F_1 hybrid by public sector
- **Pusa Manjari:** (Pusa Summer Prolific Round x Sel.-11)

- **Pusa Hybrid-3:** Fruit slighty club shaped
- NDBG-4
- Pant Sankar Lauki-1- Resistant to PM

Others

- Punjab Komal, Gutka
- Harit
- **Non Crook Neck variety-** Pusa Samridhi, Pusa Naveen, Arka Bahar

BITTER GOURD

Botanical Name : *Momordica charantia*
2x=2x=22,

Family : Cucurbitaceae

Origin : Indo-Burma

- Tendril is found
- Bitter principle momordicin
- Immature tender fruits are harvested
- Bitter gourd fruits are rich in iron
- If temperature goes above 40°C induces maleness.
- **Pant Karela-1:** Highly resistant to red pumpkin beetle
- **MDU-1:** Induced mutant

Varieties

(A)Selection

- Pusa Vishesh-Suitable for pickling and dehydration
- Pusa-Do-Mausamı, Coimbatore Long, Priyanka, Preethi, Arka Sumit
- Konkan Tara-Suitable for export
- NDB-1
- Arka Harit, $CO\text{-}_1$
- RHR BGH-1: F_1
    - ☆ Phule Green Gold- Green Long x Delhi Local
    - ☆ Sowing Time Jan-Feb: in North East India
    - ☆ Fruits ready 60-70 DAS

SUMMER SQUASH

Botanical Name : *Cucurbita pepo*
Family : Cucurbitaceae
Origin : Mexico

- Leaf Silvering is due to moisture Scarcity.
- F_2 generation can be used
- Tendril is absent
- **Harvesting stage:** at 1/3rd maturity
- It is also known as bushy cucurbit, orNamental cucurbits
- **Punjab Chappan Kaddu-1:** Resistant to DM, PM, CMV and red pumpkin beetle
- **Introduced variety-** Patty Pan, Australian Green, Early Yellow Prolific
- **Hybrid-** Pusa Alankar

SNAKE GOURD

- Botanical Name: *Trichosanthes anguina*
- Family: Cucurbitaceae
- Origin: India
- Bower system of training is best for it.
- Variety- Konkan Shweta , CO-1, CO-2,PKM-1
- MDU-1 Hybrid
- **Winter Squash**- Vilavati Kaddu
- Variety - Arka Suryamukh, Pink Banana

PUMPKIN

Botanical Name : *Cucurbita moschata*
Family : Cucurbitaceae
Origin : Mexico

- Amphidiploid (2n=4x=40) in nature
- Orissa-85% Area and 90% production
- Yerusseri Prepared from immature fruits
- ***Cucurbita ficifolia***: Fig leaf Gourd or Malabar Gourd (Perennial).
- **Varieties**

- ☆ Arka Chandan - Pleasant Aroma
- ☆ Pusa Viswas- Selection
- ☆ Arka Surya Mukhi - Resistant to fruit fly
- ☆ Pusa Vikas
- ☆ Pusa Hybrid-1
- ☆ CM-14

- ❒ Channel and hill system of cultivation is most useful for higher yield.
- ❒ Chief pollinator - Honey bees.
- ❒ Male and Female parent can be grown in 1:3 ratio.
- ❒ Flowers of Pumpkin are more nutritive than fruits.

POINTED GOURD

Botanical Name : *Trichosanthus dioca*

Family : Cucurbitaceae

Origin : India

- ❒ Dioecious, perennial, highly cross pollinated and asexually propogated vegetable
- ❒ *Salicylic acid* enhanced postharvest life of fruit
- ❒ Propagation by vine cutting 2000-2500 cttings/ha.
- ❒ Fruits are particularly recommended during convolescence for regains health after illness
- ❒ **Varieties** - Bihar Sherif, Kashi Alankar, Kalyani, Shankolia, Damodar, Swarna Alaukik, Swarna Rekha
- ❒ FP-1 (Round green) developed by ANDUAT, Ayodhya
- ❒ FP-3 (Spindle-shaped, green striped) developed by ANDUAT, Ayodhya
- ❒ FP-4 (Spindle-shaped, light green) developed by ANDUAT, Ayodhya
- ❒ **Hybrid RS**-841 is rootstock of pointed gourd
- ❒ **CHES Hybrid-1,2**
- ❒ 10% male plant is necessary to get high yield

ASH GOURD

Botanical Name : *Beninca hispida*

Family : Cucurbitaceae

Origin : Japan

- Fruits at maturity have white waxy surface
- Annual hipsid (Rough with bristle like hairs).
- Variety: Mudlier , Pusa Ujjwal, CO-1, Shakthi
- It is also known as chinese preserving melon
- Longest storage life, among cucurbits.

CHOW-CHOW

Botanical Name : *Sechium edule*

Family : Cucurbitaceae

2n=28

Origin : Mexico

- Single seeded fruit in cucurbits
- Most nutritious among cucurbits.
- Herbaceous, perennial, monoecious, climbing vine.
- High Calcium content
- Propagation: Whole mature fruits (Vivipary)

SPINE GOURD (Kakrol): Momordica Dioica

- Fruits are small, about 2.5-4 cm long, ovoid/round like **covered with soft spiness**
- Grown in tribals in MH, Gujarat, Bihar and oddisa on a limited scale
- Dioecious, perennial 2n=28
- Propagation: Tuberous roots
- Mature fruits with hard seed inside are not suitable for eating

SWEET GOURD/KHEKSA: Momordica cochinchinensis

- 2n=28
- Dioecious, perennial, climber with tuberous root
- Fruits are ovate about 10-20cm long with numerous closely set conical spine

RIDGE GOURD

Botanical Name : *Luffa acutangula* : 2n=26

Family : Cucurbitaceae

Origin : Asia

- Commercially trained or niffin system
- Used in Bathing Sponge
- **Varieties**
  - ☆ Pusa Nasdar
  - ☆ Konkan Harita
  - ☆ Punjab Sadabahar
  - ☆ *L. hermaphrodia* k/n as **Satputia** is Hermaphrodite bearing small fruit in clusters. However it is not recognized as a separate spp as it crosses readily with *Luffa acutangula* and produce fertile hybrid.
  - ☆ Kashi Khusi - Hermaphrodite variety, cluster bearing
  - ☆ Punjab Sadabahar
- Hybrid - Surekha, Arka Sujat

SPONGE GOURD

Botanical Name : *Luffa cylindrica*

Family : Cucurbitaceae

(Kashatoki) is Sanskrit Name

Origin : Asia

- Sponge Gourd contains higher protein and carotene than ridge gourd.
- Both Luffa spp. contain, a gelatinous compound called luffien
- Both Luffa spp. are used as stir-fried vegetables
- Fruits are harvested at tender stage.

Varieties

- Pusa Chikni - Selection and early variety
- Pusa Supria
- Pusa Sneha
- Phule prajakta
- Harita – F1 hybrid

Anthesis: 4-8 am (morning)

CHAPTER-7

Bulb Crops

ONION

Onion : *Allium Cepa*

: Alliaceae (Amaryllidacae)

2n=24

Origin : Asia

- The Sanskrit Name of onion is *Palandu*.
- Onion ranks second after tomato in areas.
- Acc. To Vavilov (1951) onion originated in area of Pakisthan.
- Fresh, preserved (paste and pickle) and dehydrated onion are main items of export from India.
- Onion is good source of vitamin B_1 and C and minerals like Ca and P.
- Onion contains flavonoid quercetin and kaemferal.
- Quercetin has many beneficial effects on human health, including anti-cancer activity, anti-ulcer effects, anti-allergy and cardiovascular protection.
- Allicin present in onion has antibacterial and antifungal activity.
- Red, purple and brown colours are due to anthocyanin, while yellow colour is due to quercetin.
- Onion is a cool season crop.
- Non bolting variety of onion: Early Grano and Pusa Red

Seed rate:

- ☆ 8-10 kg/ha for *Rabi*
- ☆ 12-15 kg/ha for *Kharif*
- ☆ 20-25 kg/ha for Direct sowing.

- In case of multiplier onion, 12-15 q bulblets are required for planting in one hectare onion area.
- Brown Spanish is most suitable for hilly areas.
- Flower stalk (seed stalk or scape) of onion is an apical extension of the stem without nodes and internodes.
- Inflorescence: Umbel.
- Onion fruit is a capsule, spilitting longitudinally.
- Onion is a cool season crop.
- The optimum temperature for seed germination and seedling growth is 20-25^{0}C.
- Growth declining starts at temperature higher than 27^{0}C.
- Trimming/clipping of top in case of taller seedlings are convenient to transplanting and reduces evapotranspiration loss, however it is not advisable.
- Generally, onions become ready for harvesting in 130-150 days after transplanting.
- Delayed harvesting results in splitting of bulbs and bolting.
- The purpose of drying is to remove excess moisture from the outer skin and neck region of bulb in order to reduce the infestation of disease causing organisms.
- It is also remove the excess moisture from the inner scales which is turn minimize the shrinking of bulb.
- Curing is additional process of drying to minimize the storage loss and to develop skin colour.
- For curing, usually 6-8 days are sufficient.
- Rapid drying may be done by air headed up to 38^{0}C.
- Top cutting: It is desirable to leave 1.5-2.0cm of the neck attached to the bulb.
- It helps in choaking neck and reduces the storage loss.
- Treatments before storage: Irradiation of onions with gamma rays at 6-9 Krad checks sprouting.
- Streptocycline @ 0.02% as post harvest treatment reduces losses due to bacterial rots in handling and storage.
- Klorocin @ 25kg/ha as pre-harvest treatment also reduces losses.
- The onion graded in three grades i.e. A (60-80mm), B (50-60mm) and C (35-50mm) grades according to their size.

- Only A and B bulbs should be kept in stores.
- For export, common big onions are packed in 8-25kg capacity open mesh jute bags.
- Bangalore Rose and multiplier onions are packed in 14-15kg wooden baskets for export.
- The onion can be stored for 6-9 months under cold storage at 0-2°C and 65-70% humidity with minimum losses.
- The used of gamma irradiation treatment and cold storage may be helpfulin reduction of post-harvest storage losses and regulating price fluctuation.
- Export is mainly confines from MH, AP and TN.
- In small onions, the varieties such as Agrifound Rose and Bangalore Rose grown in KN, Telangana and AP; Mutlore and Natty, the local multiplier types and also CO-4, CO-3 and Agrifound Red varieties are multiplier onion grown in Tamil Nadu are also exported.
- Bolting: Premature seed stalk formation in the crop grown for bulb production is called bolting.
- It makes the bulb light, fibrous and poor storage life.
- If bulbing and bulb development coincide with low temperature (10-12°C), bolting may take place.
- White cultivars are more sensitive to bolting.
- Sprouting and rooting of bulbs: It is one of the most important disorders in the storage and causes heavy post-harvest losses.
- Sprouting in white cultivars is reported more than in pink or purple cultivars.
- Sprouting is also associated with excessive soil moisture at maturity and excessive use of nitrogen.
- The temperature from 10-25°C increase sprouting.
- Thich necking: It refers to a failure to develop mature bulbs.
- Thick necked bulbs are difficult to dry after harvest and are more prone to sprouting and decay during storage.
- Incomplete bulbing may be due to delayed sowing and transplanting, a cool growing period or the excess use of fertilizers.
- Sjinning: Skin cracking (exposure of fleshy outer scale) may be caused either by mechanical shocks to the bulbs or by change in bulb shape associated with regrowth.
- If the skin contains <20% moisture, they are notably britle.
- The pungent oOnions have more storability than the non pungent one.

- Selfing and massing was also associated by Jones and Mann (1963) for improvement of onion.
- Dehydration ratio 10:1

Red colours varieties *(Rabi)*

- Bhima Raj, Bhima Red, Bhima Super, Bhima Kiran, Bhima Shakti, Pusa Red, Punjab Selection, Punjab Red Round,Punjab Naroya, N-2-4-1.

White colours varieties

- Pusa White Flat, Pusa White Round, Agrifound White, Udaipur-102, Punjab-48, Bhima Shweta, Bhima Shubra, Phule Safed.

Yellow coloured Onion

- Early Grano (*Kharif and Rabi*) Spanish Brown, Phule Suvarna, Arka Pitamber.

Small Common Onion

- Agrifound Rose, Arka Bindu, Bangalore Rose- **Suitable for export**

Multiplier Onion (A. cepa var. aggregatum)

- CO-1, CO-2, CO-3, Co-4, Co-5 MDU-1, Agrifound Red – These are mostly grown in Tamil Nadu throughout the year.

Kharif

- **Niphad-53** (N-53), Agrifound Dark Red, Arka Kalyan, Baswant-780, Arka Niketan (*Rabi and Kharif*), Phule Samarth (*Kharif and Rabi*)

Hybrid

- **Arka Pitamber** (White x Red)
- Arka Kirtiman, Arka Lalima, Udaipur-101, 102.

GARLIC

Garlic : *Allium sativum*

Alliaceae : 2n=16

Origin : Central Asia

- In Sanskrit garlic is known as *lasuna.*
- Garlic is a sexually sterile diploid.
- It is a herbaceous for bulb production and a biennial for seed production.
- India ranks IInd in area and production after China.
- The variety G-282 and Agrifound Parvati are the great demand for export.

- These varieties have 50-60mm bulb diameter with 10-15 cloves of bigger size.
- Allicin is the antibacterial substance of garlic.
- Frost hardy crop.
- Allin: A colourless, odourless and water-soluble amino acid present in uninjured garlic.
- On injury of the cells, an enzyme, allinase comes in contact with alliin and causes its breakdown into sulphur containing product allicin.
- The pungency of garlicis due to diallyl disulfide.
- In M.P., garlic is mainly grown.
- Garlic crop is vegetatively propogated by single cloves.
- 500-800kg cloves of 8-10mm diameter are required to one ha.
- In Lahaul-Spiti district of Himanchal Pradesh bulbs are first sown in nursery and after six weeks they are transplanted in the main field at proper spacing.
- Garlic requires medium black, fertile loamy soil with good content of potash.
- **Spiliting**- delayed harvesting or irrigation after long spell of drought.
- For past few years, rubberification problem is noticed in Rajkot of Gujarat and Nilgiri hills of Tamil Nadu and it is increasing by day by day.
- It is controlled by application of zinc sulphate and ammonium molybdate.
- The **aerial bulbils** formation is also quite common in garlic.
- G-282 is seen more prone for this type of disorder.

Cultivars

- Agrifound Parvati (G-313)- It is long day type, Big sized clove, can be grown only in northern hills.
- Yamuna Safed-3 (G-282), Agrifound White, Yamuna Safed, Yamuna Safed-2, Godavari, Shweta, Gujarat Garlic-4, Bhima Omkar, Bhima Purple.
- Boax @ 10kg/ha increase the bulb size and yield.

LEEK

Leek: *Allium porrum* syn. *A. ampeloprasum*:

Alliaceae: 2n-4x=32

Leek is a non-bulb forming member of onion family.

- Leek is biennial crop.
- Leek is grown for its blanched stem and leaves.
- Grown only in Kitchen garden in India.

- Used as salad and in flavouring the soups.
- Quercetin and kaempferol is present in leek.

Cultivars- London Flag, American Flag, Elephant, Goliath, Palam Phaustic

- PPL-1 Selection, Good replacement in green onion.
- **Hybrid-** Apollo.
- **Seed rate:** 5-6kg/ha.
- Blanching is an important operation in leek.
- White tip of leek caused by *Phytophthora porri.*

CHAPTER-8

Root Crops

RADISH

Radish: *Raphanus sativus*:

Brassicaseae (Cruciferae): 2n=2x=18

- Origin: Europe and Asia.
- Commonly known as Muli, Mula in India.
- The Name radish comes from the Latin word *radix*, meaning root.
- It can easily grow and becomes ready for use in about 4-8 weeks.
- Radish is prized an appetizer.
- It is used in preparation of soups and stuffed in paratha making.
- The mougri or rat-tail radish (*R. sativus var. caudatus*) does not produce the fleshy root but is grown for its long slender pods which are eaten raw in salad or cooked as vegetable.
- The '*Ogura*' cytoplasm of radish was transferred to cabbage and cauliflower to develop cytoplasm-gene male sterility (CMS).
- Radish is highly cross-pollinating with strong SSI.
- Self incompatibility is overcome by bud pollination by hand or high cancentration of CO_2.
- Radish seeds are good source of non-drying fatty oil suitable for soap making, illumination and edible purpose.
- Radish is good source of Vit-C (15-40mg/100g).
- Pink-skinned radish is generally richer in ascorbic acid than the white-skinned one.
- Radish root contains glucose as a major sugar.
- Pectin and pentosan are reported in radish.
- Colour of pink cultivars is due to the presence of anthocyanin.

- Portion adjacent to the root tip had higher thiocyanate than the portions near to the root top or shoulder.
- Radish is helpful in liver and gall bladder troubles.
- Roots, leaves, flowers and pods are active against G^{+} bacteria.
- *Raphanus sativus* is not known in its wild state.
- Cultivated radish was originated from *R. raphanistrum*, which is widely distributed as weed in Europe.
- The edible portion of root develops from both the primary root and hypocotyls.
- Volatile iso-thiocynates is responsible for pungency in radish.
- Thiocyanate content is known to increase with increase in sulphur supply.
- The inflorescence is typical terminal raceme of cruciferae.
- The leaves of radish are a good source for extraction of protein.
- **Flower:** White or Pink colour.
- **Fruit:** Siliqua.

Seed rate: 8-10kg.

- Radish is having sporophytic self-incompatibility system.
- The pungency increased with rise in temperature above 15°C
- Kvarta is a short duration variety suitable for protected cultivation.
- Roots of radish prepared for seed production is known as Stackling.
- Round Cherry Belle- Red coloured variety of radish.

Cultivars

Tropical or Asiatic	Feature
Pusa Desi	First variety of radish developed by IARI
Arka Nishant	Resistant to pithiness, premature bolting and root branching and forking (multiple disease resistant variety
Pusa Mridula	Red colour cultivar (also used for table purpose)
Pusa Chetki	Suitable for growing in hotter months
Japanese White	Introduction from Japan and recommended by IARI
Pusa Rashmi	Green type × Desi type
Pusa Gulabi	First entire pink fleshed nutritional rich radish variety
Pisa Jamuni	First purple fleshed unique trait nutritional rich radish variety
Palam Hriday	Pink fleshed
Punjab Safed	White-5 × Japanese White

Tropical or Asiatic	Feature
Jaunpuri Mooli	It is a local variety of Jaunpur district of UP., root,s are very long, thick and White. On an average each root wt. 2.5-3.0kg
Jaunpuri Giant	Exceptionally long, more than one metre and white in colour
Pusa Himani	(Radish Black × Japanese White) Only variety which can grown throughout the year
Rapid Red White Tipped	Very early variety, mature in 25 day, root are small, red in colour with white tip
Scarlet Globe	French Breakfast
Scared Long	-
White Icicle	-

Others: Chinese Pink, Nadauni, Kalyani White, Punjab Ageti, Punjab Pasand, Kashi Sweta, Kashi Hans,

- Akash in is a physiological disorder of radish caused by boran (B) deficiency.
- Brown heart is common in radish is due to Boron deficiency.
- Radish yellow is caused by *Fusarium oxysporum f raphani.*
- White rust is caused by *Albugo Candida.*
- Black spot caused by *Xanthomonas vesicatoria.*
- Mustard saw fly and pointed bug are the pest of radish.

Rat-tail radish (*R. sativus var. caudatus*). grown in punjab and Rajasthan for long, thin and immature pods.

- Pod of rat-tail radish contains:
- Moisture- 92.3%
- Protein- 1.3%
- Fat- 0.3%
- Fiber-optic 1.1%
- Calcium- 7.8mg.

CARROT

Carrot/ Queen Anne's Lace : *Daucus carota var. sativa:*

Apiaceae (Umbelliferae) : 2n=2x=18:

Origin : Afghanistan.

- Carrot is nature's wonderful gift to manking.
- The carrot (*Daucus carota*) gets its Name from the French word carrotte, which in comes from the Latin word carota.

- Carrots contain anthocyanin and β-carotene, which acts as an anti-oxidant.
- A well known carrot dish is Carrots julienne.
- Grated carrots are used in carrot cakes and carrot pudding.
- Carrot juice is also extracted from roots and taken as drink.
- Kanji prepared from black carrot juice is a very good appetizing drink.
- Carotene can be added to poultry feed to intensify skin and egg yolk colour.
- Sweet dishe like Halwa, Gajarpak, carrot puddings and preserve are very popular.
- Seed is source of essential oil which is used for flavouring of liqueurs and kinds of food substitutes.
- The essential oils extracted from carrot roots have anti-bacterial property.
- In India, the carrot is said to introduce from Persia.
- Present day carrot was originated from wild carrot *Daucus carota var. carota*, also known as Queen Anne's Lace.
- The fleshy tap root is designated as conical root botanically.
- It consists of primarily of phloem (cortex) and xylem (core).
- Good quality roots have maximum of cartex and minimum of core.
- Self coloured core is desirable.
- Rostte of leaves arisen from the stem.
- The inflorescence is a compound umbel consisting of many umbellets bearing small white flowers.
- Each primary ray is terminated by a secondary umbel (umbellet).
- A large primary umbel of flower stalk (King umbel) or first order umbel may contain 50 umbellets, each as many 50 flowers.
- Secondary, tertiary and quaternary umbel is produced.
- The carrot seed is a binocular fruit is called schizocarp, which at maturity splits into two in dehiscent mericarps.
- The seed germination is epigeal.
- Anthocyanin pigment is responsible for red and purple colour.
- Yellow and orange colour of carrots is the to alpha and Beta carotene.
- High sugar content, mild taste and deep orange colour in temperate carrots and red or purple attractive colour in Asiatic carrots are also associated with superior quality.

- Carrot is a rich source of high carotene content (provitamin A) and reportedly has anticarcinogenic effect due to carotene
- Inflorescence of carrot - compound umbel.
- Emasculation is laborious and time consuming.
- Carrot has abundant Beta carotene, the precursor of Vit- A and gives its characteristic orange color.
- Vitamin A activity is highest in the older cortex cells.
- Optimum temperature range for carotene/anthocyanin synthesis is 16-25°C and poor synthesis below 16°C and above 25°C.
- A long day as well as high temperatures above 25°C leads to bolting in Asiatic type carrots without adequate root formation.
- In the Northampton plains, the Asian varieties are sown from August - October, While European varieties are sown in October-November.
- **Seed rate:** 6-8kg.
- Carrot seeds contain germination inhibitor- carrotal.
- Immature seeds contain maximum level of carrotal.
- Seed germination may take place from 7th day of seed sowing and is completed in about 20 days.
- Excess nitrogen causes branching, more foliage growth compared to root, hairy and fibrous roots.
- Critical period of crop-weed competition observed between 2-8weeks after sowing.
- Mulching is advantageous in carrot for retaining soil moisture and preventing "green shoulder"
- High dry matter content is associated with better storage and handling.
- Sugars can be estimated by a refractometer.
- Sugar and flavour of carrot roots can be judged by organoleptic test also.
- High dry matter is useful for processing industry.
- Poole (1937) was first person who reported heterosis in carrot
- Thompson (1961) and Hanschke and Gabelman (1963) were the first to detect and analyze male sterility.
- The first carrot hybrids were available in early 1960s in the USA.

Varieties

Asiatic varieties	Feature
Pusa Kesar (Local Red × Nantes Half Long)	Roots stay longer in the field without showing any sign of bolting
Pusa Mighali (Pusa Kesar × Nantes)	Roots are rich in carotene content (11571mg/100g and Early sowing ,self coloured core and capable for seed setting in plains
Arka Suraj	Resistant to PM and rich source of carotene
Pusa Rudhira	It's give more than 50% higher yield than Pusa Kesar
Pusa Asita	Black colour variety
Pusa Vrishti, Pusa Vasuda, Sel-233	-
Temperate varieties	
Zeno	Introduction from Germany and suitable for Nilgiri hills of TN
Pusa Yamdagni (Sel-5) EC-9981 × Nantes	-
Ooty-1, Nantes, Nantes Half Long Early Nantes	-
Chantenay, Danvers, Imperator	-

Forking- Carrots with multiple tap roots (forks) are known as forking.

- ☆ It deteriorates root quality and salability.
- ☆ Damage to tap root of carrot seedlings also promotes forking.

- ❒ Splitting or cracking - Carrot roots are also subjected to longitudinal cracking.
  - ☆ A long dry spell followed by rain or irrigation also leads to splitting.
  - ☆ Large sized roots are more prone to splitting than small one.
- ❒ Cavity spot- deficiency of calcium
  - ☆ Cavity spot incidence also occur due to delay in harvesting .
  - ☆ In carrot root, sucrose is most abundant endogenous sugar.
- ❒ The yellow and orange carrots contain relatively more carotenoids than red or purple colour.

TURNIP

Turnip : *Brassica rapa*:

Brassicaceae : 2n=2x=20.

- ❒ In India it is popular in Jammu and Kashmir, Punjab, Haryana, Himachal Pradesh and Uttar Pradesh.
- ❒ It is cool season crop.

- It is hardy crop and can tolerate frost.
- Turnip root consisting of swollen spherical hypocotyl (napiform).
- Fruit is siliqua and called as pod.

Seed rate: 3-4 kg.

- Thinning is generally done 10-15 days after seed germination, if crop is dense.
- Turnip mosaic virus resistance (TuMV)- Single dominant gene
- Club root (Plasmadiiphora brassicae) is the major disease of turnip.
- SSI is common in turnip.

Varieties

European cultivars	Asiatic cultivars
Pusa Chandrima (Japanese White × Snowball)	Pusa Kanchan (Lo`cal Red Round × Golden Ball)
Pusa Swarnima (Japanese × Golden Ball)	Pusa Sweti- White colour
Purple Top White Globe	Punjab Safed-4
Golden Ball	-
Snowball	-
Early Milan Red Top	-

BEETROOT

Beetroot : *Beta vulgaris spp.*

Vulgaris : Chenopodiaceae: 2n=2x=18:

Origin : Mediterranean region.

- It was domisticated first as leafy vegetable (leaf beet).
- It is known as garden beet or red beet or table beet.
- The inflorescence is a large spike.
- Beet seed is multigerm seed ball.
- Fruit containing 2-6 seeds (kidney shaped and brown colour).
- It requires cold treatment (4-10C) treatment for 2 weeks or longer flower induction.
- The root meant for seed production is called stecklings.
- Beetroot was probably originated from Beta maritima (sea beet).
- Related wild species- B. atriplicifolia and B. macrocarpa.
- Colour differences (zoining pattern) between xylem and phloem tissues are due to difference in intensity of pigmentation.

- Red violet colour: β- cyanine.
- β- cyanine is a nitrogen containing compound with chemical properties similar to anthocyanin (70-90% of B- cyanine are betanine.
- Yellow colour: β- xanthin.

Beet cultivars can be classified into four categories-

Shape	Variety
i- Flat shaped	Flat Egyptian
ii- Round or globular	Detroid Dark Red and Crimson Globe
iii- Short top shaped	Crosby Egyptian, Early Wonder, Asgrow Wonder
iv- Half long	Half long blood, Winter Keeper
v- Long	Long Dork blood

Ideal temperature for growth and root development is 16-20°C

- Internal black spot (brown heart or heart rot) physiological disorder caused by boran deficiency.
- The boran deficiency is common in alkaline soil.

CHAPTER-9

Others Tuber Crops

SWEET POTATO

Sweet Potato/Irish potato/White potato: *Ipomoea batatas*:

Convolvulaceae: 2n=6x=90:

Origin: New World (South America and Mexico).

- It was first described by Linnaeus as *Convolvulus batatas.*
- Common sweet potato is a hexaploid (2n=6x=90) but most of other Ipomoea species are diploids.
- The out-crossing nature, high rate of somatic mutation and high heterozygosity.
- The tuber commonly eaten as boiled and baked.
- Fresh tuber is cooked as vegetable.
- Tuber pulp is used in preparation of gulab-jamun and other sweet dishes.
- The starchy, succulent storage roots provide raw materials for industrial used as starch and for alcohol production.
- Sweet potato provides high calorie to the diet.
- During cooking, starch to convert to moltose, which is responsible for sweet taste.
- Sweet potato is a good source of vitamin C.
- Storage root of yellow-orange fleshed cultivars contains high amount of carotenoids, whereas in white fleshed types it's quantity is very low or none.
- The protein content of most of cultivars ranges between 1.5-2.5%, which is low in sulphur containing amino acids.
- Raw sweet potatoes contain a trypsin inhibitor.
- After cooking the inhibitor is destroyed.

- Herbaceous perennial but cultivated as annual.
- It is potential stable food crop grown for its tuberous roots.
- It is grown in more than 100 countries as a source of starch, protein and carotene.
- The crop is grown in the most part of India, the leading state in the terms of area and production are Odisha, West Bengal, Uttar Pradesh, Bihar, MH, TN, MP and KN.
- The average productivity of world is 15t/ha, whereas the Indian average is very low (about 8t/ha.).
- Largest area and production of sweet potato in the world: China (85% of the total world production).
- Adventitious root are found.
- The variation in storage root dry matter and starch content among sweet potato cultivars largely depends upon the activity of ADPG (Adenosine Diphosphate Glucose pyrophosphorylase enzyme)
- High ADPG pyrophosphorylase activity occurs in cultivar with high starch content whereas, it's activity is low starch content cultivars.
- High yielding cultivars have high bulking rate over a long period
- Cultivars with intermediate and low storage root yield have a high bulking rate for short duration or low bulking rate for a longer duration.
- Based on bulking period, cultivars are classified into
  - short duration or early maturing (12-17 weeks)
  - Medium duration (17-21weeks)and
  - Long duration or late maturing (more than 21 weeks).
- Cooler night air temperature (15-25^{0}C), higher K application and 2 or 3 subsequent irrigations during the 5 to 13th weeks of growth period have reported to enhance the bulking rate of root.
- Sweet potato is generally short-day plant for flowering.
- Ideal day length for sweet potato flowering 11.5hrs.
- Highly cross pollinated crop; pollinator- honey bee and bumble bee.
- Saprophytic SI observed in sweet potato.
- Fruits are capsule: containing 4 seeds however only one or two develop.
- Seeds are hard, black in colour and require scarification or tested with concentrated sulphuric acid for 30 minute.
- Sweet potato is a warm season vegetable.

- In India sweet potato is generally cultivated as a rainfed crop.
- Sweet potato grows best at temperature more then 34^{0}C.
- Warm sunny days and cool nights are favourable for better tuber development.
- Ideal temperature for tuber formation - 20-30^{0}C.
- Night temperature less than 15^{0}C suppress storage root formation, growth and yield.
- Required light intensity for tuber formation 18,000-40,000 Lux.
- It is very susceptible to frost.
- Sweet potato is sensitive to saline and alkaline conditions.
- In tropical regions, Cl^- and SO_4^- of dodiumand calcium salt is a critical limiting factor for high productivity and expansion in cultivation of sweet potato.
- Sweet potato is commercially propagated by vine (stem) 40,000-50,000 cuttings/ha
- Cutting from shoot apex are best planting materials.
- The set of 20-50g or 2.5-4cm are generally used for propagation.
- Cut root pieces produce more vigorous shoot growth, but less uniform storage roots than vine cutting.
- Sweet potato cultivars are sensitive to photoperiod and widely differ in flowering habit.
- Problems of incompatibility and sterility also impedes in controled pollination.
- Pox and scurf is a disease severe in neutral and high pH.
- CCC and SADH are the best PGRs used for increasing tuber yield in sweet potato.
- Sweet potato matures in about 95-150 day after planting.
- The enlarged adventurous roots produced the underground tuber.
- *Ipomoea trifida* ancestor of sweet potato.
- Central Tuber Crop Research Institute (CTCRI) established in Thiruanathapuram in 1963.
- AICRP for tuber crops in 1968.
- In India the synthesized of pentaploid sweet potato (5x=75).
- Sweet potato weevil (*Cylas formicarius elegantulus*) is the major pest.
- Mass selection has been particularly valuable in development of insect pest resistance in sweet potato.

- Breeding of sweet potato for Fusarium wilt resistance is a novel work.
- Nemagold, Jewel and Pope are root knot nematode resistance.

Cultivars

Variety	Feature
Sree Bhadra	Good trap crop against root knot nematode, dry matter of 30-33%, protein 1.2-1.5% and β-carotene 800-1000 IU/100g.
H-41	Excellent cooking quality with low fibre content and susceptible to sweet potato weevil
H-42 (Velladamph × Triumph)	Excellent cooking quality, free from fibre
Rajendra Sakarkand-5 (Cross 4 × M-5)	Tolerant to waterlogging and weevil; suitable for two-tier cropping.
Rajendra Sakarkand-35	Dual purpose variety, moderately tolerance to weevil incidence and cercospora root spot
Rajendra Sakarkand-43	Suitable for late harvesting to fetch higher price
Rajendra Sakarkand-47	Suitable for late harvesting to fetch higher price
Rajendra Sakarkand-92	Good cooking quality with sweet taste
VL- Sakarkamd-6	Rich in carotene
Sree Rethna- (S-187 × Sree Vardhini)	High in β-carotene (3200-3500 IU/100g. Excellent cooking quality
Varsha	Double cross hybrid.
Gouri, Shankar	H-219 × S-73
Sree Vardhini, Sree Nandini,	

Others variety: Kalmegh, Sree Arun, Sree Varun, Samrat, CO-1, CO-2, CO-3, Kiran, Kalibnga, Gautam, Sourin, Kisan, Konkan Ashwini, Indira Madhur, Indira priya, Indira Narangi

- **Indira Naveen-** early bulking variety.
- **Indira Nandini-** it is field tolerant to weevil.
- **Narendra Shakarkand-9** -it is susceptible to drought and cracking.
- **Kamala Sundari-** rich in β-carotene and vitamin-C
- Organoleptic is the feasible means of evaluation for acceptable quality traits.
- It includes taste, feel, smell, colour and appearance.
- The cultivars White Star and Pelican Processor are develop for high starch content in the USA.
- Sweet potato breeding received special attention after the establishment of the IITA and AVRDC (now, World Vegetable Center).

CASSAVA

Cassava/Tapioco/Manioc/Mandioca/Yucca/Indioca: *Manihot esculenta* Crantz.

Euphorbiaceae: 2n=4x=36

Origin: Tropical America.

- Southwest Mexico is also considered as a region of origin or diversity.
- Fruit: Schizocarp (usually contain 3 flat hard seed).
- Cassava is photosensitive crop
- It is highly tolerant to drought
- The ideal temperature for growth is 25-30°C
- In India highest area of cassava cultivation is in Kerala.
- Productivity is highest in India in the world.
- Yellow colour of flesh is due to presence of carotene.
- Cassava is intermediate between C4 and C3 pathway plants.
- Vascular blue lining or vascular streaking is present in cassava
- The toxic principle present is cyanoide or hydrocyanic acid (HCN) or prussic acid.
- It is present in all plant parts.
- The HCN content ranges from 5-77mg in peel, 1-40mg in peeled tubers and 0.3-29mg/100g in leaves.
- Manihot species are segmental allotetraploids.
- Cassava roots are rich source of carbohydrate.
- The largest producer of cassava in the world is Nigeria, which is alone accounts 18.5% of total world production.
- Cassava is an important food in the tribal areas.
- Cassava is used as food in the form of cooked or mashed tubers.
- Cassava is also used for chips, sago and vermicilli.
- Root contains 30-40% of dry matter, which is principally carbohydrate (80-90%).
- Roots are good source of vitamin C (30-35mg/100g).
- Linamarase an enzyme that is released on the rupture of cassava cells.
- Linamarase, a naturally occurring enzyme, converts the cyanogenic glucosides, linamarin and lotaustralin to HCN.

- Most of HCN is confined to the phelloderm.
- Glucosides levels higher than 50mg/100 of edible part is considered poisonous and when it is greater than 100mg/100g, they are dangerous to the health and can cause death.
- The lethal amount of HCN for human consumption is 50-60mg.
- The consumption of about 150g unprocessed cassava high in HCN could cause death.
- Detoxification can be done by soaking tubers in water and discarding the soaked water.
- Cassava is a monoecious species.
- Flowers are protogynous.
- Pistillate flowers open 1-2 weeks before the staminate flowers.
- The pollen is large and sticky.
- It's starch is used as industrial raw material for production of alcohol, biodegradable plastic and cattle feed.
- The cassava leaves can also be utilized for indoor rearing of Erisilk worm.
- Tuberization is fast under short-photoperiod.
- It is delayed with reduction in yield under long photoperiod (more than 10-12h).
- Cassava is commercially propagated vegetatively through stem cutting (setts).
- Sweet type cassava deteriorates faster than bitter type.

Varieties:

Sree Sahya (H-2304)	Multiple crossing hybrid developed from crossing of 5 parents
Sree Harsha	Triploid cassava variety
M-4	High acceptability among the general public for its excellent culinary quality
Sree Prakash, Sree Jaya, Sree Vijaya, Nidhi	

For human consumption and starch production the ideal time for harvesting is 8-10 months.

YAM

Yam : *Dioscorea spp.*

Dioscoreaceae : 2n=4x=40, 36, 2n=6x=54

Origin : West Africa and Southeast Asia.

- Dioecious
- Yam is derivative of the tribal African word niam.
- It is also known as igName (French) and Name (Spanish).
- In edible yams, *D. alata and D. esculenta* are extensively cultivated.
- Africa alone produces about 90% of the and covers about 95% of the area.
- Alkaloid dioscorine ($C_{13}H_{19}0_2N$).
- It is generally negligible in cultivated species.
- Diosgenin is extracted from *D. viarum, D. floribunda, D. composita, D. deltoidea.*
- Two temperate spp. *D.opposita and D. japonica*
- Fruit is a dehiscent trilocular capsule with upto six seeds.
- Mulching in yam is very useful operation.
- Most effective training is 'pandal' system.
- Yam is a major stable food crop of Africa.
- Only seven species are cultivated for human food Namely- *D. rotundata, D. cayenensis, D. dumetorum, D. trifida, D. alata, D. esculenta, D. bulbifera.*
- Generally yams are harvested 7-9 month after planting.
- Double harvesting is done in yam.
- Harvesting is done in December-January.
- Yam tubers under go a period of dormancy for 2-3 months.
- Treatment with 4-8% solution of ethylene chlorohydrin breaks the dormancy of yam tubers.
- Soaking of tubers in 1% solution of maleic hydrazide (MH) delayed the sprouting in storage.
- Average yield of white yam is 18-25t/ha.
- Vines of yam require staking.
- Main sapogenin present is diosgenin which is starting point of several corticosteroid drugs.
- Poisonous (Insecticide) type-*D. dumetorum and D. hispida.*

- A 30g minisett is rapid for seed yam production.

Cultivars

Yam	Varieties
Greater yam	Sree Keerthi Sree Roopa Sree Shilpa (Da- 140 × Sree Keerthi) - intervarietal hybrid Agua Forastero Purple Ceylon, Florida
Lesser yam	Sree Latha Sree Kala Konkan Kanchan
White yam	- Sree Subhra Sree Priya Sree Dhanya

Sree Latha- is the 1st cultivar of an edible yam released in India.

ELEPHANT FOOD YAM

Elephant Food Yam/Suran/Zamikand: *Amorphophallus paeoniifolius* (syn. *A. campanulatus*)

Araceae (aroid family): 2n=2x=28:

Origin: Tropical Asia.

- EFY is a cheap source of carbohydrate (24% starch) and rich in minerals and vitamin A and B
- Corm is used for preparation of curries, gulab jamun and pickles.
- Petioles are also used after boiling just like asparagus.
- Rich source of starch (24%).
- Starch can be extracted and made into gelatinous noodles.
- Glucomannan is made from its flour.
- Most of the EFY species contain calcium oxalate crystals.
- The substance is toxic to human, and if eaten raw causes irritation of mouth, tounge and throat.
- Precious gifts of nature.
- Tuber juice extract is used to cure high blood pressure.
- It's seeds are also used to counter irritants and as tonic and laxative.

- About 14 species occur in India.
- Only 3 species, Namely; *A. paeoniifolius, A. kanjak (syn. A. rivieri) and A. variabilis* are edible.
- Flowers are greenish violet coloured.
- Flowers are extremely protogynous.
- EFY, the genetic variability is very narrow due to erratic flowering behaviour.

Varieties

Gajendra	High yielding selection from Kovvur region of AP. And good cooking quality and totally free from acridity
Santragacchi	Rough tubers slight acridity
Sree Padma	Selection from local cultivars from Shabad region of Kerala
Narendra Ravi (NDA-5)	Mid season variety free from acridity
Narendra Asha (NDA-9)	Early variety free from acridity
Kovvur	-

Harvesting in about 180-200 days after planting; however it can harvested between 6-9 months after planting.

- Paddy straw is used for mulching.

COLOCASIA

Colocasia/Taro/Dasheen/Arvi/Cocoyam/Eddoe: Colocasia esculenta

Araceae (Aroidae): 2n=2x=28 (Dasheen) 2n=3x=42 (eddoe)

Origin: India

- Fermented acidic product called 'poi'.
- *Colacasia* flour is good baby food.
- A decoction of the leaves with the scraped root of yasi yasi (*Syzygium effusum*) is drunk to treat stomach disorders.
- The young leaves of toro are rich in Vitamin C, and the roots are rich in a starch composed of amylose (28%) and amylopectin (72%).
- *Colacasia esculenta var. esculenta* the dasheen (banda) types having big corm (mother corm) with none or few developed cormels (side tubers).
- *Colacasia esculenta var. antiqourum* (*syn. var. globulifera*), the eddoe (Arvi) type, which have comparatively smaller corms and large number of cormels.
- Adventitious roots develop from lower portions of the corm.
- *Colacasia* has no dormancy.

- Metsbure in colocasia due to Ca deficiency.

Cultivars

Sree Rasmi	Cormels, corms, and leaves are free from acridity
Narendra Arvi-1 and Narendra Arvi-2	Both corm have less acridity and grown in both summer and rainy season
Satmukhi	-
Sree Pallavi	-
Narendra Banda-2,3,21	-

The eddoe type colocasia mature in 150-180 days and dasheen type 180 - 200days.

CHINESE POTATO

Chinese potato/Coleus/koorka/ratala: *Solenostemon rotundifolius* (syn. *Coleus parviflorus*) Labiatae

Origin: Tropical Africa

- Tuber used as vegetable.
- Sree Dhara is the variety of coleus.
- Propogated by stem cuttings.
- Suckers and tubers can also be used for planting.

Tannia: *Xanthosoma sagisttifolius*: Araceae: Origin: West Indies in Central America.

- It is used for making roll of leaves.
- Propagated by corms and cormels.

Yam bean/mishrikand/sakalu: *Pachyrrhizus erosus*: Papilionaceae (Fabaceae)

- **Origin :** Tropical America
- **Rajendra Mishrikand-**1, give higher yield of 40-55t/ha.

CHAPTER-10

Pea and Beans

PEA

- **Pea**/Vegetable pea/Garden pea/Green pea: *Pisum sativum var. hortense*: Fabaceae (Papilionaceous):2n=2x=16: Origin: Central Asia:
- Garden pea occupies a prominence position in the agricultural economy of the country.
- Acc. to Blixt (1970), the Mediterranean is the primary center of diversity with secondary centers in Ethiopia and the Near East.
- The genus Pisun comprises small number of taxa.
- All taxa within Pisum are diploid.
- Pea is an annual herbaceous plant.
- Uttar Pradesh first ranks in area and production followed by Bihar and Madhya Pradesh.
- Pea are highly nutritive and contain a high percentage of digestible protein, and appreciable amount of carbohydrates and Vit A and C.
- The inflorescence is raceme arising from the axil of a leaf.
- The flowers are typical papilionaceous with green calyx comprising of five united sepals, five petals (one standard, two wings and two keels).
- The stamens are in diadelphous (9+1) condition.
- Nine filaments are fused to form a staminal tube, while the tenth is free throughout it's length.
- The gynoecium is monocarpellary, with o ovules (up to 13) alternately attached to the two placentas.
- Pea is strictly self-pollination in nature.
- Outcrossing is negligible (less than 1%).
- Outcrossing as high as 60% is reported in Peru.

- The symbiotic association with Rhizobium in roots also found.
- Inflorescence is recemose.
- Flowers are hermaohrodite (Perfect).
- Fruit- pod or legume.
- Endocarp present in most of cultivars except those of sugar pea.
- The pea seed consisting of two large cotyledons surrounding the embryo and enclosed by the seed coat (testa).
- The shelling percentage varies between 30-56% in different cultivars.
- The germination in pea is hypogeal.
- Optimum soil temperature for seed germination is 20-24°C.
- Seed rate:
  - For dwarf var. 125-150 kg/ha.
  - Medium tall- 80-100kg/ha.
- Peas are sown in Rabi season from beginning of October to the end of November in plains.
- In northern plains, the best time of sowing is 20th October to 20th November.
- Edible quality is usually determined by using a tendrometer
- Tendrometer readings, when adjusted for temperature, correlate well with edible quality.
- Tendrometer as judged by sweetness, tenderness and starchiness.
- The amount of alcohal-insouble solids in seed is another method for quality determination.
- During maturity, sugar content decrease rapidly and increase in starch and other polysaccharides.
- Conversion of sugar to starch is governed by temperature.
- The high CO_2 content within the pod provides micro-environment.
- The pods get freeze at 10°C.

Cultivars

Early Groups	Features	Early Groups	Features
Arkel	Introduce from England (UK).	Jawahar Mater-3	T_{19} x Little Marvel
Pusa Pragati	Resistant to DM	Jawahar Mater-3	T_{19} x Little Marvel

Early Groups	Features	Early Groups	Features
Asauji	Collected from Amritsar by IARI New Delhi	Hara Bona	Developed from pureline selection from local field pea variety
Early Badger	-	Meteor	-
Little Marvel	(Chelsea Gem × Suttoons Alaska)	Harbhajan (EC-33866)	Extra early variety
Hisar Harit (PH-1)		Ageta-(E6)	
PM-2	(Early Badger × JP-3).	Early December	
Kashi Nandini		Kashi Uday	
Kashi Mukti		VL Ageti mater-7	
Mid -Groups			
Bonneville	Introduced from USA	Perfection New Line	
Kashi Shakti	(Hara Bone × NDVP-8)	P-35	
Azad Pea-1		JM-1	
Pant Uphar		Punjab-88	(Pusa-2 × Morassis-55)
NDVP-8		NDVP-10	
Ooty-1			
Late Groups			
Multi-freezer		Kashi Samrath	
Kashi Samridhi		NP-29	
Lincoln	Introduced from USA		

Edible Podded Pea

- ☆ Sylviya, JP-19, UN-53, Madhu, Arka Sampoorna, Organ Sugar.

- ❒ **Hara Bona-** Flowering starts 21 days after sowing and first harvesting in about 50 days after sowing.
- ❒ **Aparna-** First dwarf variety
- ❒ **Meteor-** Introduce from England
- ❒ **Bonneville-** Introduce from USA.
- ❒ Powdery mildew of pea is governed by a single recessive gene (*er-1and er-2*).

Diseases of Pea

Diseases	Casual organisn	Inheritance	Resistant variety
Powdery mildew	*Erysiphae polygoni*	Single recessive (er-1,er-2)	Arka Ajit, Arka Karthik, JP-83, JP-71, PRS-4
Ascochyta foot rot	*Ascochyta pisi*	Single dominate gene	
Wilt	*Fusarium oxysporium pv. pisi*	Single dominate gene (Fw)	
Rust	*Uromuces pisi*	Single dominate gene	

Multiple pests and diseases resistance cultivars-

- JP-179, JP-9, JP-501, JP-501, JP-Batri Brown-3, 4.

FRENCH BEAN

French bean or Snap bean	: *Phaseolus vulgaris*:
Fabaceae	: 2n=2x=22:
Origin	: Southern Mexico and Central America.

- The bean grown for immature pods (vegetable) is called snap bean.
- Edible Portion 94% of pod
- Staking is important operation in pole type varieties
- Snap bean also known as garden bean, salad bean and green bean.
- Common bean and navy bean are applicable only to pulse type, where dry seeds are used.
- The snap beans were earlier known as string beans, because of fibrous thickening of the vascular bundles in pods.
- Snap bean is an excellent source of protein, vitamins A and B and minerals.
- Snap bean contains good quality protein.
- It has anti-nutritional factors viz., flatulence causing factor.
- Beans are also said to be anti-diabetic and good far bladder burns.
- It acts as Carminative, depuratitive, diuretic and emollient.
- The key character of the genus is coiling of corolla in the form of keel.
- There are five cultivated species of Phaseolus Namely,
- French bean or snap bean - P. *vulgaris*, runner or scarlet bean- P. *coccineus,* tepary bean P. *acutifolius var.* latifolius, Lima, sieve,butter or Madagascar bean P. *lunatus* and P. *polyanthus.*
- P. *coccineus* is generally cross-pollinated and other four species are self-pollinated with only small degree of crass- pollination.
- Snap bean is an erect or twining annual herb, polymorphic for growth habit, seed characteristics and maturity.

Seed rate:

- ☆ Pole type-25-30 kg/ha
- ☆ Bush type-65 kg/ha

Varieties of French bean

Pole type	Features	Bush type	Features
Kentuky wonder	Introduced from USA	Pusa Parvati	X-rays mutant of wax pod
Tweet wonder	Resistant to Anthracnose	Arka Komal	Introduced from Australia
Pusa Himlata		Contender	Tolerant to PM and mosaic
Blue Lake		Arka Suvidha	Photoinsensitive, resistant to rust
TKD-1		Arka Anoop	Photoinsensitive, resistant to Bacterial wilt and rust
SVM-1		Arka Sharath	Photoinsensitive
Lakshmi		Arka Bold	Photoinsensitive, resistant to Bacterial wilt

Most of French bean varieties are day neutral except semi-pole (shart day plant)

- It is sensitive to both water excess and water stress condition
- Blossom drop and ovule abortion of common problem at high temperature 35°C
- Seed moisture below 12% enhances the cracking of cotyledon.
- Transverse cotyledon cracking (TCC) is major physiological disorder in french bean
- Hypocotyl cracking or necrosis disorder is the germination disorder in french bean
- Hypocotyl cracking or necrosis disorder due to low calcium content in seed

There are four main growth habits Namely;

- Determinate upright type
- Indeterminate upright type
- Indeterminate,weak stemmed, non climbing or semi climbing type
- Indeterminate, weak stemmed, climbing type

COWPEA

Cowpea: Southern pea, Black eye pea, Beans (West Africa), Asparagus bean, Yard long bean : *Vigna unguiculata* :

Fabaceae (Leguminosae) : 2n = 2x = 22

Origin: Southern Asia and India

- In India, cowpea is also known as "*lobia*".

- The Name lobia is probably of Greek origin derived from the word *lobos* meaning pod.
- In Sanskrit, lobia meaning *alluring*.
- The Name cowpea was first used in America.
- Earlier English Name of cowpea was Cavalance, and was first time used in West Indies.
- It is very important legume vegetable grown in almost all the states of India.
- Its tender green pods are cooked alone or mixed with other vegetables.
- Pods are rich source of protein.
- In tropical Africa and Latin America, cowpea is primarily used in the form of dry seeds cooked as a pulse in various dishes, and in Asian countries, it is cultivated both for green pods used as vegetables and dry seeds and immature seeds.
- It is also frozen and canned.
- Cowpea seed is also used as baking powder for biscuits, vegetable milk and legume starch.
- Cowpea is rich in protein. Protein concentration ranges from 3-4% in green leaf.
- Carbohydrate content of dry seed is more than 50%.
- Amino acids viz., lysine, leucine and phenylalanine contents are relatively high in cowpea.
- It has anti-nutritional factors *i.e.,* a-amylase inhibitor.
- The cultivated cowpea is polymorphic; however, there are three major forms of cowpea which are collectively considered cowpeas. These are -
  - ☆ common cowpea, *Vigna unguiculata* cultigroup *unguiculata*
  - ☆ yard long bean, *Vigna unguiculata* cultigroup *sesquipedalis* and
  - ☆ catjang cowpea, *Vigna unguiculata* cultigroup *cylindrica.*
- The yard long bean is the most liked cowpea for vegetables purpose.
- The term "cultigroups" is more appropriate than subspecies for these cowpeas, because genetic differences are small.
- Stems are climbing require support.
- Flowers start to appear as early as 4-6 weeks after seedling emergence and edible pods are formed in about 2 weeks after anthesis.
- Harvest most often begins about 70 days after planting and may continue for 25-30 days, Pods are 30-80cm in length.

- Common cowpea is bushy in nature.
- Cowpea is an annual, bushy or vine plant.
- Flower is polypetalous with irregular corolla.
- Both, the androecium and gynoecium are enclosed inside the keel tube.
- The flowers of cowpea are hermaphrodite and self-pollinated.
- Pollen viability of cowpea is high.
- Dehiscence of anthers is much earlier to anthesis and therefore, the cowpea is strictly self-pollinated in nature.
- Male sterility has also been reported in cowpea.
- High night temperature can cause male sterility in cowpea.
- Germination of seed is epigeal.
- Weight of 100 seed is about 22g.
- Cowpea can be grown at the temperature range of 20°-35°C.
- If temperature increases above 36°C pod set is adversely affected.
- Likewise, if temperature goes below 15°C, the growth, flowering and fruiting are drastically reduced.
- Growth is optimum at 27°-30°C day and 17°-22°C night temperature.
- For bushy and erect growing cultivars 20-25 kg seeds and for climbing (pole type) cultivars 10-12 kg seeds per hectare are sufficient.
- Rhizobium treated seeds ensure adequate nitrogen supply to the crop from atmosphere.
- Sowing time in northern India, there are two sowing times of cowpea:
    - **Summer season:** Seed sowing starts in February and sowing may continue up to April.
    - **Rainy season:** Seeds are sown from June-August.
- In southern India, places like Bangalore where climate is mild, it can be grown almost throughout the year.
- Green pods are ready for picking 10-12 days after flower opening.
- Over-mature pods loose their quality due to more fibre formation.
- Picking should be done at 3-4 days interval for getting good quality tender green pods with high yield.
- The average green pods yield varies from 125-150 q/ha.

Varieties

Determinate varieties (Bush Type)	Feature
Pusa Rituraj	**Photo-thermo-insensitive**, dual purpose variety (both green pod and dry seeds are used).
Pusa Phalguni	This cultivar was selected from Canadian cultivar 'Dolique due Tonkin' and it suitable for spring sowing.
Pusa Barsati	Introduction from Philippines
Pusa Komal	Indeterminate, photo-insensitive and resistant to bacterial blight
Arka Samrudhi	
Arka Suman	
Kashi Shyamal	Tolerant to golden mosaic virus and *Pseudocercospora cruenta*
Kashi Gauri(IIHR Sel. 16 x Sel. 2-1)	bush type and dwarf, photo-insensitive and early suitable for sowing in both spring-summer and rainy season. it is tolerant to golden mosaic virus and *Pseudocercospora cruenta*
Kashi Unnati (Cowpea- 263 x Arka Garima)	Photo-insensitive
Kashi Kanchan- (Sel. 2-2 x BC-244002 x Sel. 2-2)	Develop from back crass
Kashi Sudha, Kashi Nidhi, Cowpea-263, Sel. 2-2, S-203, Narendra Lobia-1,2	
Indeterminate varieties	
Arka Garima	
Philippines Early	
Anaswara	

CLUSTER BEAN

Cluster Bean (Guar) : *Cyamopsis tetragonoloba*:

Family : Fabaceae : 2n = 2x =14

Origin : West Africa

- Among beans, cluster bean commands a premium place.
- Its tender green pods are used as vegetable.
- Rajasthan accounts for nearly 70% of the total production of cluster bean in country.

- The guar gum is major industrial product.
- The major world suppliers of guar gum are India, Pakistan and USA.
- India accounts for 80% of the total guar production in the world.
- Being a drought tolerant, this vegetable can be grown successfully as a rainfed crop, and quite suitable as an intercrop.
- The genus Cyamopsis consists 3 species, out of which *C. tetragonoloba* is the only species of economic importance.
- Ancestor of cluster bean : C. *senegalensis* (African species).
- The immature, tender green pods are used as vegetable.
- Green pods are dehydrated and stored for use during off-season as vegetable.
- Flour is also prepared from its seeds, which is in great demand in paper textile industries and as food.
- Guar seed contains 19 to 43% mucilaginous galactomannan gum, which is used as a thickener in food products *viz.*, ice cream, bakery mixes and salad dressings.
- Used in textile, paper, cosmetic, oil industries, and is also useful want for explosives.
- In industry, gum works as a thickener, flocculants, binder, lubricant and filtrate.
- The cluster bean crop is also used as nutritive fodder for the cattle.
- Green pods are good protein (3.7%), vitamin A (830 IU) and vitamin C (49 mg) per 100g edible portion.
- Flowers are axillary raceme (borne in cluster).
- Flower: white or pink in colour.
- White seed is preferred for gum extraction in comparison to black seed due to better quality of gum.
- The preferred seed size for gum extraction is 4mm diameter.
- Guar galactomannan gum is contained in the endosperm of the seed, which lies in a spherical layer beneath the seed coat.
- The protein content of guar seed ranges from 27 to 37%.
- Guar seed contains 4% edible oil, which has 64% oleic acid.
- Varieties with low gum content are selected for vegetable.
- Guar is a warm season crop, hardy, drought tolerant and adapted well in arid and semi-arid climates.
- Sensitive to frost and temperature below 15°C affects germination.

- Optimum soil temperature for seed germination is 30°C.
- The best temperature range for pod setting is 15-25°C, and al very high temperatures during rainy season, there are flowers and pods drop.
- Guar prefers long days for growth and short days for flowering.
- In northern India, two crops are grown i.e. (i) rainy season crop- seeds are sown in June-July, and (ii) summer season crop- seeds are sown in March-April.
- 15-20 or 30-40 kg of seeds is enough for on hectare sowing.
- Zinc deficiency characterized by interveinal chlorosis, leaf thickening, defoliation and general plant stunting.
- Therefore, soil application of 20-25 kg zinc sulphate per hectare is recommended for Zn deficient soils.
- Kharif / rainy season crop is harvested in October-November and summer erop is harvested May-June.
- Picking should be done before pods become stringy and seeds become large enough to cause the pods bulge around the seed.
- First picking is done after 2-3 weeks of flowering or bout 50-60 days after sowing.
- Picking of pods is done by twisting or cutting from the plants.
- Green pod yield varies from 15-20 tones/hectare.

Cultivars

- **Pusa Mausami-** First pickling about 80 days after sowing. Suitable for sowing only rainy season.
- **Pusa Sadabahar-** Suitable for sowing in summer and rainy season.
- **Pusa Navbahar (Pusa Mausami x Pusa Sadabahar)-** Pod picking starts 40 days of sowing.
- **Durga Bahar (Pusa Navbahar x ROC 401)-** Photo-insensitive.
- **P-28-1-1 – (Pusa Navbahar x IC 11521) -** It is photo-insensitive, highly branched and yields 140-150 q/ha.
- **Pardeshi-** grown in Gujarat.
- **Punjab-**4, Sharad Bahar, Durga Bahar,
- **Durgapura Safed-** multiple use variety.
- Triploid and teraploid mutants from the colchicine treatment have been reported.

BROAD BEAN/ PULSE OF POOR MAN

B. N. : *Vicia fava*

Fabaceae : 2n=2x=12, 14

Origin : Madagascar

- Broad bean is also known as "Faba bean, horse bean, field bean,Windsor bean, tick bean and bakla bean.
- Mainly cultivated in Latin America.
- It is cross pollinated crop and pollination by Insects.
- An illness "favism"caused by it.
- Broad bean contains a growth-dressing factor other than trypsin inhibitor, haemagglutinin or favism- inducing factor.
- Favism is an inherited disease (genetic disorder) in which a person lacks an enzyme called glucose-6-phoasphate dehydrogenase.
- Inhalation of the pollen or ingestion of the seeds may incite the condition known as favism, a severe haemolytic anemia in children.
- Levodopa an important chemical found in broad bean is used in the treatment of Parkinson's disease.
- Dancing style of stigma present in broad bean.
- The entire plant including, leaves, stems, pods and immature bean contains levodopa.
- All beans are susceptible to frost except Broad bean.
- Among the beans, broad bean is the only bean which is grown as winter crop.
- It can be withstanding cold temperature as low as 4°C.
- The large-seeded form is known as broad bean, *Vicia faba var. major.*
- The small-seeded types are, *Vicia faba var. equina and Vicia faba var. minor.*
- *Vicia faba var. paucijuga* which is not commonly grown.
- Seed rate: 70-100kg/ha.

Varieties:

- Pusa Sumeet and Pusa Udit.

LIMA BEAN

Lima bean : Butter bean, Double bean

B.N. : *Phaseolus lunatus*:

Fabaceae : 2n=2x=22

Origin : Central America (Mexico and Guatemala).

- Often crass pollinated crop.
- The largest area and production of lima bean in world: USA.
- It is used to prepare wine.
- It is cross compatible with French bean.
- Lima bean is not closely related to French bean and Runner bean.
- A fertile segmental allotetraploid (4x=44) was produced from crass between *Phaseolus lunatus x P. polystachyus.*

Seed rate:

- **Small seeded:** 20-35 kg/ha.
- **Large-seeded:** 45-62 kg/ha.

Cultivars:

- **Pole type-** Florida Butter, Challenger, Carolina Butter and King of Garden
- **Semi-pole type-** Hopi and Wilbur.
- **Bush type-** Baby Potato, Baby Fordhook, Burpee, and Henderson
- Improved variety 'Kodaikanal-1' It has bolder seeds with matured seed yield of 1200kg /ha.

INDIAN BEAN/ FIELD BEAN/ HYACINTH BEAN

B.N : *Lablab purpureus* .Syn. *Dolichos lablab, D. purpureus*).

Family : Fabaceae (Leguminosae) 2n =2x= 20, 22, 24

Origin : India

- Indian bean is also known as hyacinth bean, **dolichos bean**, bonavist bean, lubia bea Indian butter bean, **sem bean**, field bean, Egyptian kidney bean, Australian pea and waby sa bean.
- Immature tender pods are used as vegetable.
- It provides a considerable amount of protein and minerals in the diet, particularly in rural population of India.

- Indian bean grows well in all tropical and subtropical regions, and major growing states are Madhya Pradesh, Uttar Pradesh, Tamil Nadu, Andhra Pradesh and Gujarat.
- As a legume crop it also fixes nitrogen in the soil.
- **Yerdcort (1970)** placed Indian bean in a separate genus Lablab from Dolichos.
- He classified L. perpureus into the following sub-species:
- *L. purpureus* subsp. *perpureus*- It includes common cultivated races of Indian bean.
- *L. purpureus* subsp. u*ncinatus*- It has more slender inflorescence with smaller pods, about 4cm long and 1.5cm wide.
- *L. purpureus* subsp. *bengalensis*- pods are linear, similar to kidney bean in appearance and quite dissimilar to those of other Indian beans.
- The green immature pods and tender seeds are used as cooked vegetables and dried seeds as pulse.
- Green shelled seeds are mostly preferred in South India.
- White seed cultivars contain low and non-toxic amounts of cyanogenic glucoside and trypsin inhibitor, whereas dark seed cultivars have high evel of both.
- Partial dominance with duplicate type of epistasis for pod yield per plant has been reported. Indian bean is a highly self-pollinated crop: however cross-pollination through insects has also been reported.
- The protruded stigma (extrovert) in a few varieties leads to allogamy.
- Hebbal Avare-1 has been evolved by involving Mutation breeding Camma irradiation (24 krad) has been used to develop varieties of Indian bean.

Indeterminate/pole Type Variety

- **Deepaliwal-** This is developed at Punjabrao Krishi Vidyapeeth, Akola, Maharashtra. This variety is a high yielding with crop duration of about 90 to 95 days.
- **Pusa Early Prolific-** Early variety and grown throughout the year
- **Pusa Sem-2** - cultivation in northern India.
- **Others:** Pusa Sem-3, Rajani, DL-453, DL-2342, DL-695, HD-1, HD-18, HD-60

Determinate and Photo-Insensitive Varieties

- **Arka Jay-** Developed through backcross and pedigree selection involving the parents Hehbal Avare and IIHR 93. It is tolerant to low moisture stress.

- **Arka Vijay-** Developed through backeross and pedigree selection involving the parents Hebbal Avare and IIHR 93, Plants are dwarf, bushy, erect. It is photo-insensitive variety and flowers in about 60 days after sowing. Flowers are white: pods short, dark green with characteristic aroma. Pods are vegetable type without parchment, It is tolerant to low moisture stress.
- **Hebbal Avare-3 (** Hebbal Avare-1 x US-67-13) Released for Karnataka.
- **CO-1**. It is a pureline selection from TNAU, Coimbatore
- **CO-2** (CO-8 x C0-1)

Others: Wal Konkan-1, VAL-125-36, Konkan Bushan.

- Rewa variety having 25.11% protein.
- **Generally pole type is photo-**insensitive in nature.
- **Suitable for dual purpose:** Arka Vijay and Konkan Bhusan.
- Bush type benas, photo-insensitive, tolerant to heat and drought: Arka Jay and Arka Vijay.
- **Suitable for vegetable purpose:** Arka Jay.

Disease of Common Bean

Disease	Casual organism	Vectors
Anthracnose	Colletotrichum lindemuthianum	
Web blight	Rhizoctonia solanii	
Angular leaf spot	Isariopsis griseola	
Floury leaf spot	Ramularia phaseoli	
Yellow flecks	MLOs	White fly
Phyllody	MLOs	Leaf hoppers
Golden mosaic	Virus	White fly

WINGED BEAN

Winged Bean/ Multiflorus bean: *Psophocarpus tetragonolobus Sy Tetragonolobus purpureus*

Fabaceae:2n=20

Origin: Papua New Guinea, Madagascar and India.

- It is also known as Gao bean, Manila bean, Four-angled bean, Princess pea and Asparagus pea.
- It is also known as – **"Soya's rival" "God-sent vegetable"**
- Winged bean is a valuable vegetable.

- All parts of the plant are edible *i.e.,* pod, seeds, tender leaves, inflorescence and tubers roots, therefore, it is said to be "a supermarket on a stalk".
- Among vegetables, winged bean is one of the richest sources of protein and oil.
- Its oil is rich in unsaturated fatty acids.
- The dry seeds contain 30-40% protein and 15-20% oil.
- The tubers have 11-15% protein.
- The leaves, pods and tubers have medicinal uses.
- A wild species, *P. palmetsorum* (2n=18) occurs in Tropical Africa.
- Trypsin and chymotrypsin inhibitor are well known anti-nutritional factors in winged bean.
- **Other anti-nutritional factors are:** amylase inhibitor, phytohemagglutinins, cynogenic glycosides and saponins.
- The winged bean seed inhibitor activity can be safely eliminated only by steaming the seeds in an autoclave at 130°C for 10 minute.
- Short day plant.
- Commercially grown in North India.
- Winged bean is propogated by seeds.

Cultivars

- **Tpt-3:** Suitable for home garden.
- **GRWB-24:** Highest pod yield and GRWB-25 gave highest root yield.
- Gui Ai- is natural dwarf mutant cultivar.

SCARLET RUNNER BEAN

Scarlet Runner Bean: ***Phaseolus coccineus***: Fabaceae

- Native to Mexico and Central America.
- It also known as Multiflora bean.
- It is long day plant and grown on limited scale in India.
- It is cross pollinated bean.
- Mature dried seeds are rich in protein (17%), carbohydrates (65%) and contain 2% edible oil.
- Pinching is practiced to reduce vine length and encourage side branches.

JACK BEAN

Jack Bean/Horse bean: *Canvalia ensiformis*: **Fabaceae:**

- Native to Central America and West Indies.
- The jack bean and sword bean are very similar to each others.
- Tender young pod pods and immature seeds are used as vegetable.
- It can tolerate drought due to its deep-rooted habit.
- It is self-pollinated crop.
- Flowers are rose to violet in colour.
- Pods are sword shaped.
- Jack bean contains a thermolabile toxic principle causing haemorrhage of the mucous lining of the stomach and intestines.
- It propagated by seeds.

SWORD BEAN

B.N.: *Canvalia gladiate,*

Family: Fabaceae

- Native to Tropical Asia and Africa.
- Immature pods and seeds are used as a vegetable.
- Flower and tender leaves are also sometime used as vegetable.
- It has a twining vine habit.
- It is self-pollinated crop.
- Flower colour: Pinkish white to white.
- The pods are sword shaped.
- Pods are curved and have strongly developed ridges than those of jack beans.

VELVET BEAN

B.N.: *Mucuna prurita,*

Family: fabaceae

- Native to Southern China and eastern India
- Velvet bean is known as *'Kapi Kacchu'* in Sanskrit.
- Cowitch and Cowage are the common English Names.
- Young tender pod used as vegetable.

- The roasted ground seeds are used as substitute of coffee in Brazil.
- The alcohol extracts of the leaves have been reported to have significant analgesic, antipyretic and anti-inflamatory effects.
- The paste of leaves is applied to ulcers.
- The mature seeds are highly toxic.
- They contain di-hydroxy-phenylalanine or dopa, glutathione, lecithin, gallic acid and glucoside.

CHAPTER-11

Leafy Vegetable

AMARANTHS

(A) Warm season vegetable

Amaranths: Poor man's leafy vegetable/ *Amarnthus spp* .

Amaranthaceae : 2n=32 or 34

Origin: India

- There are leaf amaranths as well as grain amaranths.
- The leaf amaranths are used for vegetable but the young plants with tender stems and leaf of grain amaranths are also eaten as leafy vegetable.
- The leaf amaranths are rich source of protein, minerals and vitamins A (9200 IU) and C (99mg).
- The amaranthus have antinutrient factors, like oxalates and nitrates.
- The oxalate result in kidney stone and nitrates from methaemoglobin in the blood.
- The genus Amaranthus is classified into two sections, Namely amaranthus and Blitopsis.
- All species for grains are under Amaranthus and leafamaranths under Blitopsis.
- The species A. tricolor mostly used as leaf amaranth has also some forms with orNamental leaves.
- C4 plants.
- *A. hypochondriacus, A. caudatus A. hybridus* 2n= 32.
- *A. tricolor, Blitopsis* 2n=34
- *A. Dubius* 2n= 4x=64.
- Primarily used as pot herb.
- Type of fruit: Glomerule

- Fruit: Utricle.
- Basically warm season crop (most common vegetable grown in summer and rainy season).
- Susceptible to water logging.
- It can be grown in soil pH as high as 10.
- **Clipping** is done in Amaranthus.
- OrNamental species *A. tricolor* bedding plants, *A. caudatus* "Love Lies Bleeding" plants with drooping rope-like inflorescence.
- Short day species *A. caudatus (syn A. paniculatus) A. cruentus (syn A. edulis)*
- Main cultivated species in india *A. tricolor*
- Most amaranths species are monoecious. Except (A. spinosus, A. dubius) dioecious.
- The grain amaranths are paleo-allotetraploids.
- Popped seeds are used for making sweet balls (*laddoo*) and cakes (*patty*).
- The basic unit of inflorescence is glomerule.
- The glomerule is crowded to form complex inflorescence, called spikes.
- The percentage of male flowers per glomerule is 0.5-1 in grain types and 10-25 in leafy types.
- Ideal temp. for seed germination is 25-30^{0}C and growth is 20-30^{0}C
- The seed rate is 2kg/ha in case of badi chaulai and about 3kg/ha for chhoti chaulai.
- Seed rate: 2kg/ha for direct seeding and 1kg/ha for transplanted crop.
- Seeds are sown at a depth of about 1.0-1.5cm.
- Amaranths is a anemophilus (wind pollinated) crop.
- Monoecious species of amaranths are chiefly self-pollinated (Blitopsis) Grain type favour cross-pollination (protogynous).
- For good leaf yield, 2-3 cutting are practiced.
- The first cutting can be made 3-4 weeks after sowing.
- Leaf yield varies from 100-200q/ha.
- Seed yield 200kg/ha.

Cultivars: *A. tricolor var.* Pusa Badi Chaulai, Pusa Keerti, Pusa Kiran, Pusa Lal chaulai,CO-2, CO-5, Arun (Photo-insensitive) Mohini.

- ☆ *A.blitum var.* Pusa Chhoti Chaulai.
- ☆ *dubius var.* CO-1

Arka Arunima – leaves contain low antinutrient factors like oxalate and nitrates.

- It is resistant to white rust under epiphytotic conditions.
- Yield 26-28t/ha.
- Arka Suguna- moderate resistant to white rust.
- Pusa Kiran (*A.tricolor × A. tristis*)
- CO-5 -double coloured leaves green and pink)

BASELLA

Malabar spinach/**Indian spinach/Malabar nightshade**/climbing vine/Poi/ Basella: *Basella spp.*

Origin: South Asia (India)

Family: Basellaceae 2n=44, 48

- The basella is a short day plant with C_4 photosynthetic pathway.
- Succulent leaves with petiole and tender stem, used as leafy vegetable.
- A food colouring dye is extracted from the berries of Basella.
- Cooked roots are used in the treatment of diarrhea.
- The leaves are rich in protein (1.6g), calcium (106mg), iron (1.6mg) and vitamin A (3250 I.U) and C (86mg).
- Basella is commercially propogated by seeds (12-15kg/ha).
- *Basella alba*- 2n=2x=48, white colour flowered form.
- Basella rubra- 2n=2x=44, red type.
- Induced polyploids with 4x=88 chromosomes were produced in *Basella and B. rubra*.
- *B. carninifolia, B. cordiifolia*- have heart shaped leaves.
- Basella mainly of two type:
    - ☆ One with reddish petioles and stems.
    - ☆ Other with green leaves, petioles and stems.

NEWZEALAND SPINACH

Newzealand spinach: *Tetragonia tetragonoides* (syn. *T. expansa)*, Tetragonaceae: 2n=2x=32

Origin: New Zealand.

- Self-pollinated and annual.
- The flavour of its leaf similar to spinach.

- It is warm season vegetable and can tolerate to high temperatures (35^{0}C).
- Leaf and stem both are used.
- In India, specially grown in the eastern hills around Shillong, Kalimpong, Darjeeling and other hill stations during summer.
- Harvesting usually beging about 40-50 days after planting.

(B). Winter Leafy Vegetable

PALAK

Spinach Beet/Indian spinach/beet leaf/palak: *Beta vulgaris* L. var. *bengalensis* (*syn B. v. var. orientalis*):

Family: Chenopodiaceae: 2n=2x=18

Origin: Indo-Chinese

- Herbaceous annual (for consumption) and biennial for seed production.
- Rich source of vitamin-A (9770 IU), Vit C-(70mg) and Ca (380mg/100g).
- Vitamin-C is present in sufficient quantity, but it destroyed easily in processing and cooking.
- The major carotenoids are lutein (1505mg) and β-carotene (1095mg/100g)
- As a palak is leafy vegetable it requireds more N_2 for crown growth, apply N @20-30kg/ha after every cutting as top dressing.
- Washing of leaves is not advocated because may take place at center of bundle.
- Palak leaves contain low oxalic acid.
- Palak contains 9.5% Oxalate.
- It can be grown on alkaline to saline sodic soils having pH 10.5.
- It is mostly grown in UP, Bihar, WB, Punjab, MP, Gujrat and MH, it is not so common in southern states.
- Leaves of this crop might have been first used in Bengal and hence it is known as *Beta vulgaris* L. var. *bengalensis*
- It is closely related to beet root and sugar beet (*B. vulgaris spp. cicla*).
- Sea beet (*B. vulgaris spp. maritima*), is the ancestor of palak beet root and some other cultivars.
- Palak produces hermaphrodite flowers.
- Crop is a highly cross pollinated and pollination by wind (Anemophius).
- Spinach have unilocular perygynous ovary.
- Production of more than 1 seed in single boll/fruit is called multigerm (2-3 seeds/fruit).

- Single fruits of palak contain 2-3 seeds.
- **Seed rate:** 25-30kg for summer crop and 10-15kg/ha for winter crop.
- Leaves become ready for first cutting in about 3-4 weeks after sowing; 6-8 cutting can be taken in winter.
- The no. of cutting is reduced with increase in temperature above 25^0C, and above 30^0C only one cutting is obtained.
- Cutting is done leaving about 5cm petiole attached to the compressed stem, for rejuvenation.
- Average yield 10-15t/ha of leaves is obtained.
- Average seed yield 600kg/ha.
- Seed weight about 8g/1000 seeds).
- Leaves can be stored for 10-14 days at 0^0C and 90-95%RH.

Cultivars

Pusa Jyoti	Polyploid variety
Jobner Green	Mutant variety
All Green	Late bolting
Pusa Harit	Sugar beet × Local palak
Pusa Palak	
Punjab Green, Banerjee's Giant, Pusa Bharati	
Arka Suguna and Ooty-1	

Pant Composite-1, Arka Anupama tolerate to Cercospora leaf spot.

Difference Between Spinach Beet and Spinach-

S.N.	Attributes	Spinach beet	Spinach
1.	B.N	*B. vulgaris var.bengalensis*	*Spinacia oleracea*
2.	Choromosome	2x=18	2x=12
3.	Sex forms	Bisexual	Dioecious, Monoecious, hermophrodite
4.	Leaves	Have entire margin	Have lobed margin
5.	Seed	Multigerms	Monogerm
6.	Tolerate	High temperature	Purely cool season

SPINACH

Spinach/Vilayati palak: *Spinacia oleracea* L.:

Chenopodiaceae: 2n=2x=12:

Origin: South West Asia (Iran/Persia).

- The word spinach actually came from a Spanish word *Hispania*.
- *Spina* in Latin means spiny fruit and *oleracea* means herbaceous herb.
- It is extensively cultivated in Europe and North America.
- It is not familiar in our country except in hilly areas because it is substituted by spinach beet, a similar type of vegetable which is a much common vegetable grown in North India.
- The wild relative of spinach is S. tetrandra, a possible ancestor and *S. turkestanica*, which are found North western India and Nepal.
- Dioecious, annual and herbaceous plant.
- It can tolerate frost better than most cool season vegetables.
- Rich in vitamin A (9300 IU) and C (28mg/100g).
- Spinach is cross-pollinating and wind pollinated.
- It is highly nutritious among leafy or salad vegetables.
- Sex expression in spinach is tetramarphic (Extreme male, Vegetative male, Monoecious, Female plant).
- Inflorescence- Apetalous staminate.
- Flower- small and greenish.
- Monogerm single seeded fruit (utricle).
- Parthenocarpic fruit often occurs.
- Seed rate- 25-30kg/ha.
- The optimum temperature for seed germination is 10-15^{0}C
- Test weight- 10g.

Cultivars:

- Virginia Savoy- prickly seeded blistered large dark green with round tip and plants are upright and vigorous in growth.
- Early Smooth Leaf- smooth seeded and small light green leaves with a pointed apex.
- Average cutting: 3-4.

- Yield: 15-20t/ha.
- Average seed yield: 700-1000kg/ha
- Janick and Stevenson (1954) interpreted the hetrogametic nature of sex in spinach-
- XX- Pistillate.
- XY- Staminate (may or may not bear seed).
- YY- Staminate (dose not bear seeds).

FENUGREEK

Fenugreek/Common Methi: *Trigonella foenum-graecum* L. :

Fabaceae: 2n=2x=16:

Origin: Southeast Europe and West Asia (between Iran and Northern India)

There of two species of economic importance-

- Common methi- *T. foenum-graecum*
- Kasuri/champa methi- *T. corniculata.*

Different between common methi and Kasuri methi-

Characters	Commom methi	Kasuri/champa methi
Flower	White	Yellow/orange
Growth	Fast	Slow
Growth habit	Upright/erect	Rosette
Seed rate	25-30kg/ha	18-20kg/ha
Maturity	Early	Late
Leaf yield	90-100q/ha	More than common methi
Seed yield	15-18q/ha	7-8q/ha

Fenugreek is the third seed spice in India (after coriander and cumin).

- Rajasthan is the largest producer of methi.
- It is considered as 'fenugreek bowl' of the country.
- More than 80% area and production of the country is contributed by Rajasthan alone.
- India is the major producer and exporter of fenugreek.
- Fenugreek seeds are rich in essential amino acid and trigonelline (150mg/100g).
- Seeds contain diosgenin (0.40-1.26%) which is useful as starting material in synthesis of sex hormones as oral contraceptives.

- Fenugreek has antidiabetic action due to presence of polysaccharide galactomanon.

Cultivars:

Pusa Early Bunching	Dual purpose, quick growing with upright shoot and bold seeds
Lam sel-1	Dual purpose, Early maturing and tolerant to root rot, PM, aphid and caterpillars
CO-1	Quick growing and dual purpose variety, field resistant to root rot
CO-2	Suitable for both rabi and kharif, tolerant to Rhizoctonia root rot
Pusa Kasuri	Small seeded variety mainly cultivated for leaves
Rajendra Kranti	Suitable for inter-cropping, both rabi and kharif season, Resistant to Cercospora leaf spot.
Rmt-1	Plant vigorous, pinkish base, quick early growth, Moderately resistant to root rot. Tolerant to PM, and root knot nematode and yield 1560-2500kg/ha.
Rmt-143	Pureline selection. It is suitable for growing in Rajasthan
Rmt-303	Determinate variety
Rmt-305	Developed through mutation breeding in Rmt-1. determinate, multi-poded. Early maturing variety. It is resistant to powdery mildew and root knot nematode.

Others: Hisar Sonali, Hisar Mukta, Hisar Suvarna, Hisar Madhavi, Gujarat Methi, Pant Ragini,

CHAPTER-12

Salad Vegetable Crops

Salad vegetable crops:Lettuce, Celery, Parsley, endive, chicory, chervil, cress, water cress.

- **Endive:** Cichorium endive
- **Chicory:** Cichorium intybus
- **Chervil:** Anthriscum cerefolium
- **Cress:** Lipidium sativum
- **Water cress:** Nasturium officinale

LETTUCE

- **Lettuce**: *Lactuca sativa* L.: Asteraceae (Compositae): 2n=2x=18: Origin: Mediterrranean region.
- The Name Lettuce is derived from latin word *lactic* (Milk).
- Vitamin A- 1650IU/100g and Ca: 50mg/100g.
- Lettuce is the most common cool season salad vegetable.
- Lettuce is very sensitive to high soil acidity.
- It is grown commercially in USA, Japan, Australia, most of the countries in Europe and South America.
- It is a annual and polymorphic plant.
- Highly self-pollinated crop.
- It contains lacteal fluid (lactucin alkaloid), which give a slightly bitter taste.
- Consumption of lettuce reduces blood cholesterol, obesity, diabetes and improved blood structure.
- Lettuce also known to be sedative, diuretic and expectorant.
- Single seeded dry fruit called an achene.
- Seeds are prone to shattering.

- Mulching is most essential.
- Lettuce seed have thermo-dormancy.
- The dormancy may be broken by exposure of seed to chilling at 4-6^{0}C for 2-3 days.
- Lettuce is very sensitive to high soil acidity.
- Fluid drilling is one of the methods to overcome seed dormancy in lettuce
- Optimum temperature for seed germination is 24-25^{0}C.
- Bolting: Lettuce is very prone to premature flowering, if exposed to warm weather (above 25^{0}C).
- Long days may induce seed stalk formation and flowering.
- Seed rate: 400-500g. test weight: 1g.
- Avg. yield: 15-20t/ha.
- There are four generally recognized morphological forms of lettuce; however stem and latin lettuce are other forms which also grown sporadically.

Lettuce type and cultivars:

- Crisp head or iceberg or head lettuce: **Great Lakes**, **Imperial 859**
- Leaf or loose leaf or bunching lettuce: **Punjab Lettuce No.1**, **Slobolt**, **Chinese Yellow**
- Butter head or cabbage lettuce: **White Boston, Big Boston, May King.**
- Cos or romaine: **Dark Green, Eiffel Tower.**
- Stem or celers or asparagus lettuce: celtuce
- Latin lettuce: **Gallega**. It is first source of lettuce mosaic resistance.
  - ☆ The most commonly grown lettuce is crisphead.
  - ☆ Progenitor of cultivated lettuce: Lactuca serricola. It was used as a bridge species in successful crosses with cultivated species.
  - ☆ Indian lettuce: *L. indica*.
  - ☆ Inflorescence is head or capitulum.
  - ☆ Recommended CO_2 in lettuce cultivation: 1000-1500ppm.
  - ☆ ID-BRI-cultivar resistant to Sulfonylirea herbicide.

Pest and Diseases:

- Downy mildew (*Bremia lactucae*) is a major disease of lettuce.
- Lettuce mosaic virus disease is seed born and transmitted by aphid.

- 'Vanguard 75', the first crisp head lettuce with resistance to LMV.
- Big vein of lettuce is due to virus (seed born), which is transmitted by fungus (*Olpidium brassicae*).
- Most important disease of head lettuce: Slimy soft rot.
- Tip burn is a physiological disorder of lettuce due to deficiency of Ca.
- Russet spotting disorder of lettuce due to ethylene injury.

CELERY

- **Celery**: *Apium graveolens L.* var. *dulce*: Apiaceae (Umbelliferae): 2n=2x=22: Origin: Eastern Mediterranean basin.
- Edible part: leafstalks or fleshy petiole.
- Its very good for rheumatism.
- Globally celery is the second most importance salad crop, after lettuce.
- In India, it is grown as a seed crop during winter.
- It was introduced in India during early 1930s.
- The crop is commercially grown in Amritsar, Gurudaspur and Jalandhar district of Punjab and to a small extent in Saharanpur district of U.P.
- The leaf stalks and petioles of celery are eaten raw as salad and used in soups.
- Celery is also processed by pickling, canning and dehydration.
- Its seeds are used as condiment in European countries and spice in India.
- Apiin (apigenin 7-apiosylglucoside) is the main flavouring glucoside of celery leaves and celeriac root.

Three cultivated species

- *Apium graveolens* L. var.*dulce* (stalk celery) leaf petiole are the edible part. It can be further divided into self-blanching and green types.
- *Apium graveolens L. var.cecalinum* (smallage or leaf celery)- whole are used as a spice.
- *Apium graveolens L. var. rapaceum* (root or knob celery)- swollen base of the stem and part of root.
- Annual for vegetable and biennial for seed production (seed produce only in hills).
- Leaves grow in rosette.
- Inflorescence: compound umbel with secondary division known as umbellate.
- Bolting and flowering take place at low temperature (10^{0}C or less).
- High temperature causes bolting bitterness in celery leaves.

- Moisture loving plant.
- Bolting and flowering simultaneous phenomena, but in celeriac, flowering (sessile flowers) may occur without bolting.
- Crop is insect pollinated due to protandry.
- **Fruit:** cremocarp (schizocarp).
- **Seed rate:** 125g needed to raise seedling.
- Weight of 2500 seed is 1g.
- Ideal temperature for seed germination is 15-20^{0}C
- Oxidiargyl is a good pre-emergence herbicide for transplanted celery crop.
- Cultivars: two types of varieties; self-blanched or yellow and green leaved.
- Green leaved var. are preferred in the Indian market.
- Self-blanched: Florida Golden and Golden White Plum.
- Green leaved: Wright Grove Giant, Ford Hook Emperor, Standard Bearer, Boston Market, Utah and Fascal.
- In India, Supreme Golden (self-blanched) and Golden No.15 var. are mostly grown.
- IARI recommended variety: Standard Bearer and Wright Grove Giant.
- Wrapping of leaf petioles with dark brown paper around them in celery is known as blanching.
- Blanching is done to make the crop crips, reduced acrid flavor, increase good flavour and tenderness.

Physiological disorders

Pysiological disorders	causes	Remady
Black heart	Ca deficiency	Spray Ca. chloride (5-10kg/ha)
Cracked stem	Bo deficiency	Spray 0.03% boron.
Pencil strip	Excess of phosphorus	Recommended doses of phosphorus
Chlorosis	Mg deficiency	Spray $MgSO_4$ @ 20kg/ha.
pithiness	Excess 'N'	Avoid heavy doses of 'N' and irrigation.

CHINESE CABBAGE

Chinese Cabbage : *Brassica spp.*:

Brassicaceae

(old Crucifereae) : 2n=2x=20:

Origin : China

- It is grown on commercial scale in China, Taiwan, Korea and Japan.
- Rich of vitamin B_1, B_2 and carotene.
- Citric acid is the major acid found in Chinese cabbage.
- Average glucosinolates is 0.198 in heading type and 0.534g kg^1 FW in leaf type.
- Pak-choi *i.e.,* **leafy type** (*B. rapa* subsp. *chinensis* or *B. chinensis.*
- **Cultivars**: Chinese Sarson-1, Palampur Green, **Pusa Sag** (Wong Box (Suttons) × Turnip), Solan Selection
- Pe-tsai *i.e.,* **heading type** (*B. rapa* subsp. *Pekinesis* or *B. pekinensis*)
- **Cultivars:** Chihili, Michihili, Wong Box, Che-foo, Spring Giant, Spring Giant, Tokyo Giant, Tropical Pride, Tropical Delight, Tip Top, China King and Oriented King.
- Hybrid var. **Optiko**
- Asveg-1, an F1-hybrid from AVRDC can be grown in tropical and sub-tropical region.
- Zhengbai No.4, has multiple disease resistance.
- Both pak-choi and pe-tsai are cross pollinated annuals with presence of sporophytic Self incompatibility.
- Male sterility of CMS type is also reported in Chinese cabbage.
- Chinese cabbage is not a popular vegetable in India. It is grown on a limited scale near big cities or kitchen gardens.
- Temperature above 25^{0}C delay head formation and reduce quality.
- Direct seeding is more commonly followed in leafy type.

Seed rate

- For transplanting 500g/ha.
- For Direct sowing 2.5kg/ha.
- Avg. yield 20-25t/ha. Hybrid- up to 100t/ha.
- Thining (5-6 leaf stage) is very important operation in chinese cabbage.
- Mulching is recommended in Chinese cabbage.
- Sensitivity to high temperature.
- 'Tipburn' is one of the most common physiological disorders of Chinese cabbage due to Ca-deficiency.
- Low O_2 (2%) and low CO_2 (2%) improves shelf life of Chinese cabbage.

PARSLEY

Parsley : *Petroselinum crispum var. cripsum* (syn. *P. hortense, P. sativum*)

Apiaceae : 2n=2x=22:

Origin : Europe.

- It is cross-pollinated crop.
- Rich in pro-vitamin A and C (90mg/100g).
- Parsley greens are used for flavouring salad, soups and coocked vegetable.
- It is also dehydrated.
- *P. crispum var. tuberosum* with turnip root parsley, both root and foliage are used in preparation of soups and stews.
- IARI recommended variety: Moss Curled and Humburg.
- **Others:** Forest green, Moss curled, Sherwood (double or triple curled), Italian Dark Green Plain Leaf, Plain or Single, Extra Tripled Curled and Imperial.
- Parsley is propogated by seeds.
- Seed rate: 250-300g.
- Seed germinate in 7-20 or more days.
- Harvesting starts about 90 days after seed sowing.

GARDEN SORREL (Khatta Palak)

Garden Sorrel(*Rumex acetosa*),

Family: Polygonaceae

- Sorrel derived from French word *"surele"* means sour
- It is close relative to rhubarb.
- Monoecious.
- Edible part: large fleshy leaf stalk (midrib).
- The term 'dock' has been used in Britain to include all members of the family Polygonaceae.
- Owing to unpleasant flavour, it is sometimes called sour dock and sour grass.
- Cultivared sorrel is two kinds-
- *R. acetosa* (garden sorrel) and *R. scutatus* (French sorrel)
- Seed rate: 8-10kg/ha.

KARAM SAG

B.N.: *Brassico oleracia var.acephala*: Cruciferae

Chakwal

B.N.: *Atripex hortensis*: Chenopodiaceae:

- Commonly known as Orach or Mountain spinach, strong flavour then spinach

CHAPER-13

Perennial Vegetable Crops

Perennial Vegetable: Asparagus, Drumstic, Chekkurmanis, Bread fruit, Globe artichoke, Rhubarb, Curry leaf, Ceylon spinach, Jerusalem artichoke and Horse radish.

ASPARAGUS

Asparagus : *Asparagus officinalis* L. *var. altilis*:

Liliaceae : 2n=2x=20:

Origin : Europe and Asia.

- Asparagus is a cool season crop.
- Edible part: spear
- Fruit: berry (1-3 seed/fruit).
- Tender shoots called 'spear' are used as vegetable.
- The optimum soil pH is 6.5-7.5 for its good plant growth
- Asparagus plants are hardy herbaceous, d**ioecious** and perennial.
- Plants established in about 1-3 years.
- Once established, it can produce up to about 25 years.
- It is good source of vitamin A (700IU).
- The main active components of this plants are mono and dicaffeoyguinic acid.
- It has high content of PUFA (Poly unsaturated fatty acids).
- Asparagus shoots are used to cure jaundice and to clean the bowel.
- Asparagus contains rutin, a compound that useful in hemorrhage (bleeding) and also diuretic properties.
- Cladophylls' is the chief photosynthetic organs.
- Carbohydrate reserves in the root also play an important role in the growth of spears.

- Tender shoots of asparagus contain a white crystalline substance is called 'asparagine'.
- Green asparagus are more popular and mainly used for fresh market.
- Light green or white are used for processing.
- Blanched (whitish) spears are preferred for canning.
- The thick receptacle of flower bud is known as heart which is used for canning.
- Blanching is practiced to blanch the young spears and get white asparagus for canning.
- Green spears have more nutritive value than the white (blanched spears).
- Closely related species of asparagus: Smilax (OrNamental asparagus).

Others Related or Namental Asparagus

A. *Chochinchinensis* with edible tubers and *A. racemosus* having edible roots.

- **Flower colour:** Whitish green (bell-shaped).
- Asparagus is propogated by crown and seed.
- Crown are traditional propogating materialand are widely used.
- One-year-old crown is used for propogation.

Seed rate

- For Direct sowing 2-3kg/ha.
- For seedling (hybrid) 1kg/ha.
- Seedling 10-12 weeks old are transplanted.
- The F1 hybrids may be all male or dioecious.
- Optimum temperature for seed germination is 20-30^{0}C.
- Optimum temperature for spear production is 16-24^{0}C
- In asparagus male plants (MM) are higher yielding than the female plnats (mm).
- Male flower are more conspicuous than the female flowers.
- Exotic varieties are grown in India.
- **Cultivars:** Perfection, Mary Washington, Marth Washington, Waltham Washington, New Jersey Improved, Jersey Queen, Jersey Giant, Reading Giant, Limbrass 22, Limbrass 126 and Palmetto.
- Recommended variety of IARI: Perfection (an exotic variety).
- These California hybrids include, Altas, Apollo, Grande and new purple asparagus called Viola or Purple passion.

- Viola has a beautiful deep purple colour when raw
- It is also sweeter than green.
- The spears can be stored at 2^{0}C temp. and 95% RH in good condition for 10-14 days.
- Asparagus rust (*Puccinia asparagi*) is most serious disease.
- Resistant cultivars are Jersey Giant and UC-157

MORINGA OR DRUMSTICK

- *Maringa oleifera*
- Moriga also known as Drumstick and it is a fast growing multipurpose tree.
- It is also known as horse radish on ben oil tree.
- Seeds of *Maringa oleifera* contains 8.9% oil.
- Flowers yellowish - white and fragrant.
- Optimum temperature for growth is 25-35°C
- Flowers are protandrous.
- Pollinator Carpenter bees
- **Cultivars -** Jaffna, Chavakacheri Murungai, Chem Murungai, PKM-1, PKM-2.
- PKM-2 is a hybrid cultivar developed at TNAU, India.
- In West Bengal, 2 distinct genotypes Sajna and Najan are cultivated.
- Moringa tree is propagated by seeds and by limb cuttings.
- Annual types are propagated by seeds and perennial types are propagated by limb cuttings.

CHEKRAMANIS

- *Sauropus androgynus.*
- Native - India and Burma.
- It is also known as "21st century vegetable" or "Multivitamin Green"
- Family - Euphorbiaceae.
- Cross pollinated and entomophilous.
- Flowers are monoecious and protogynous.
- Grown for its leaves and tender shoots.
- Propagation is by seeds and stem cuttings

BEAD FRUIT

- *Artocarpus altilis:* Family – Moraceae: 2n = 56.
- The plant is monoecious.
- **Variety -** Yellow Heart.
- Stimulative parthenocarpy is found in bread fruit.
- **Propagation:** By root cutting

GLOBE ARTICHOKE

B.N.: *Cynara scolymus*: Fmaily-Compositae

- Edible part: flower bud or flower head
- Native: Mediterranean region
- Propagation: Sucker
- The main active components of this plants are mono and dicaffeoyguinic acid.
- It has a high content of PUFA (Poly unsaturated fatty acids).
- The thick receptacle of flower bud is known as heart which is used for canning.
- **Cultivars**- Green Globe, Purple Globe, Bull, Spinosa Sarda, Imperial Star.
- Plants grown from seeds come into bearing in about 8 months while plants raised through suckers flower earlier.

RHUBARB

Rheum rhaponticum.

Family: *Polygonaceae*

2n=4x = 44.

- Edible part - large thick leaf stalk or petiole.
- Propagated by divisions of crown.
- The optimum temperature for proper development of stalks is 15.6°C
- **Variety:** Strewberry, Madonald, Valentine, Victria, Cherry Red

CHOW-CHOW (PERENNIAL)

- Single seeded fruit belonging to the family cucurbitaceae.
- Monoecious.
- The fruit is vivipary in nature
- Single seeded

- Richest source of nutrient among gourds.

CURRY LEAF

Murraya Koenigi.

- Self-Pollinated.
- A volatile oil, a crystalline glucoside "Ksenigin" from leaves and "Murayin" from the flowers are found.
- Fruit - berry.
- Propagation is by seeds.
- Ployembryony has been reported in curry leaf.
- Cultivars DWD-1 and DWD-2.

JERUSALEM ARTICHOKE

Helianthus tuberoses.

- Family-compositae.
- Edible part-tuber.
- It is a commercial source of levulose, used as a sweeting agent by diabetic patients.
- Glucofractons principle carbohydrate in tubers.

SEAKALE

- Family - Cruciferae.
- Grown for it's young leaves and shoots

HORSE RADISH

- Pungency is due to the presence of antallyl isothiocynate and butyl thiocynate.

CEYLON SPINACH OR WATER LEAF

- *Talinum triangulare.*
- **Native-** Brazil.
- The crop is highly self pollinated.
- Propagated through seeds or herbaceous stem cuttings.

CHAPTER-14

Spices and Condiments

Spice Research Institutes

- Indian Institute of Spices Research (IISR), Calicut, Kerala
- National Research Center for Seed Spices (NRCSS), Ajmer, Rajasthan
- Indian Cardamom Research Institute (ICRI), Idukki, Kerala
- Spices Board, Kochin, Kerala

Classification Spices and Condiment

Sr. N.	Family	Common Name	Botanical Name	Origin	2n
1.	Piperaceae	Black pepper	*Piper nigrum*	Indo-burma	128
2.	Zingibraceae	Small cardamom	*Elettaria cardamomum*	India	48
		Ginger	*Ginger officinalis*	South east Asia	22
		Turmeric	*Curcuma longa*	-do-	62
3.	Lauraceae	Cinnamon	*Cinnamomum verum*	Srilanka	24
4.	Myrsticaceae	Nutmeg	*Myristica fragrans*	Indonesia	42
5.	Myrtaceae	Clove	*Syzygium aromaticum*	-do-	44
6.	Myrtacene	All spice	*Pimenta dioca*	West Indies	22
7.	Solanceae	Capsicum	*Capsicum annum*	S. America	24
8.	Rutaceace	Curry leaf	*Murraya koenigi*	-	18
9.	Apiaceae	Asafoetida	*Ferulla foetida*	-	
		Dill	*Anethum graveolens*	Eurasia	
		Coriander	*Coriandrum sativum*	Mediterranean region	22
		Cumin	*Cuminum cyminum*	-do-	14
		Fennel	*Foeniculum vulgare*	-do-	-
		Aniseed	*Pimpinella anisum*	-do-	-
		Bishop's weed		-do-	-
10.	Iridaceae	Saffron	*Crocus sativus*	India	
11.	Araceae	Sweet flag	*Acorus calamus*		
12.	Orchidaceae	Vanilla	*Vanilla planifolia*	America	32

Sr. N.	Family	Common Name	Botanical Name	Origin	2n
13.	Fabaceae Leguminaceae	Fenugreek	*Trigonella foenum graceum*	Europe	16
		Tamarind	*Tamarindus Indica*	T. Africa	24
14.	Guttiferac	Kokum	*Garcinia indica*	India	24

Spices and its chemical content

Plants	Chemical content	Essential oil (%)
Black pepper	Piperine ($C_{17}H_{19}O_3N$)	1-3
Small cardamom	Cineol-terpenyl acetate	-
Ginger	Zingerol	1.9-2.2
Turmeric	Curcumin	.24
Cinnamon	Cinamaldehyde	.5-2
Cassia	Cinamaldehyde	
Nutmeg	Tuebein, Sabinene	-
Clove	Eugenol	-
All spice	Phenol	
Capsicum	Solanine	
Curry leaf	Koenigine	
Asafoetida	Ferumine	
Dill	Carvone	**1.5-4.0**
Coriander	Linalool	.1-1
Cumin	Cuminol	2.5-3.5
Black cumin	Carvone	
Fennel	Limone	.7-1.2
Aniseed	Anithol	
Bishop's weed	Thynol	
Saffron	Cicrocrocin	.5-1
Sweet flag	Acorin	
Vanilla	Vanillin	
Fenugreek	Diosgenin	
Kokum	B hydroxyl acetic acid	

Plants	Part used
Black pepper	Fruit, seed
Small cardamom	Fruit, seed
Ginger	Rhizome
Turmeric	Rhizome
Cinnamon	Bark
Nutmeg	Aril

Plants	Part used
Clove	Unopened flower bud
All spice	Fruit, seed
Capsicum	Fruit
Curry leaf	Leaves
Asafoetida	Oleoresin gum
Dill	Fruit
Coriander	Leaves, seed
Cumin	Fruit
Fennel	Fruit
Aniseed	Fruit
Bishop's weed	Fruit
Saffron	**Stigma**
Sweet flag	Rhizome
Vanilla	Fruit
Fenugreek	Seed
Kokum	Fruit

- **Dioecious-** Nutmeg, All spice, Kokum, Juniper
  - ☆ India's share in world trade of spices is around 18%.
  - ☆ Spice oil is obtained by steam distillation method.
  - ☆ India is known as home of spices.
  - ☆ Largest area under spices - Rajasthan.
  - ☆ Highest production under spices -Andhra Pradesh,
  - ☆ Black pepper and turmeric constitute more than 50% of total export earnings among spices.
  - ☆ USA-major importer & consumer of leoresins (50%)

Plants	Varieties
Black pepper	Karimunda, Sreekara, Subhakara, Pooenima, Panchami, Mudigree-1, Mudigree-2 Hybrids- Panniyur-1, Panniyur-3 PLD-2 –High oleoresin content (15.5%)
Small cardamom	Mysore type, Malabar type, PV-2, Vazukha type: Cross between Mysore × Malabar Mudigree-1 is the best suited for HDP (1x1m)
Large cardamom	Bebo, Golsey, Ramsey, Ramla, Seremna Sikkim-1,2
Ginger	Rio deJanerio (Introduced)-most popular *var.* Surbhi , Surari –Mutant Suruchi, Suphrabha Himeiri- tolerant to rhizome rot (best for green ginger)

Plants	**Varieties**
Turmeric	Rajendra Sonia, Pratibha, Krishna, Roma, Sudarshana, Saguna, Suvarna, Prabha, Kranti, Rashmi, Srobha, Sugandham, Suroma Allepy turmeric- Best in the world BSR-1&2 and CO-1 (mutant)
Cinnamon	Konkan Tej, Navashree, Nithyashree, YCD-1 Sugandhini, PPI(C)-1
Nutmeg	Konkan Sugandha- Hermaphrodite variety Konkan Shrimanti, Konkan Swad
Clove	Pevang, Lenzipore, Amulva
All spice	Jamaican Allspice, Mexican Allspice, Honduran Allspice.
Capsicum	
Curry leaf	Suwasani
Asafoetida	
Dill	
Coriander	Pant Harithna, Sindhu, Sadhana, Rajendra Swathi , Swathi RCr-41 Karan -small seeded variety
Cumin	RZ-19, GC-1, GC-2, RZ-209, Vijapur-5
Fennel	RF-101, RF-125, RF-35
Aniseed	
Bishop's weed	
Saffron	
Sweet flag	
Vanilla	RMT
Fenugreek	Rajendra Kranti, Hissar Sonali, Lam Sel.-1
Kokum	Konkan Amrit

Plants	**Propagation**
Black pepper	Shoot cutting
Small cardamom	Sucker
Large cardamom	Suckers
Ginger	Rhizome
Cinnamon	Seed
Fenugreek	Seed
Coriander	Seed
Cumin	Seed
Curry leaf	Seed
Clove	Seed (mother of clove)
Vanilla	Stem cutting

1. Black pepper / Black gold

- Oil of Black pepper mainly Consist of Terpense
- King of spices
- Multifunctional spice
- Climbing evergreen plant
- Inflorescence- Catkin
- The pH of 4.5-6.0 is ideal
- Fruit- Single seeded berry
- Majorities of cultivated varieties are monoecious
- Largest producer - Vietnam
- India accounts for 54% of total area and 26.6% of total production in the world
- **Fruiting branches :** Plagiotropes
- **Top shoots:** Orthotropes
- **Hanging shoots:** Geotropes
- Western Ghats are best suited for cultivation
- Pepper Research Station, penniyur established in 1949
- Harvesting-Nov-Feb in plains; Jan-March in Hills
- Kerala accounts 96% of total production, KN (3.5%), TN and Pondicherry
- Tripod-stand for harvesting
- North America - major importer of Indian pepper.

2. Small cardamom

- Most expansive spice in the world
- Queen of spices
- Fruit- trilocular capsule
- Kerala (60%) leading producer of cardamom
- Harvesting : Oct-Nov.
- Summer showers are essential for flowering
- Inflorescence- Panicle
- *Pentalonia nigronervosa* (banana aphid) is a vector of katte or mosaic disease
- The first commercial level production of plantlets in spice crops through tissue culture was achieved in cardamom

3. Large cardamom /Nepal cardamom / Greater Indian cardamom

- *Amomum subulatum*
- Sikkim the largest producer of large cardamom
- Shade loving plant (seophyte)
- Foorkey and Chirke are the two viral disease

4. Ginger

- Posses only Aroma not flavur
- Jamaica and India produced the best quality ginger in the world
- India accounts 70% of total world production and 50% in world export.
- Kerala leading producer of ginger (60% area and 25% production)
- Cochin ginger/Indian ginger best in the world.

5. Turmeric

- India accounts 76% of total world production
- AP leading turmeric producer.
- *C. aromatica* are tetraploids and set seed
- Lakadong turmeric got GI tag in Meghalaya
- Curcumin – 6.8-7.5%
- Turmeric is antioxidant due to phenolic character of curcumin.

6. Clove

- The term “clove” is drived from french word ‘colv’/clou meaning nail
- Indonesia accounts for 68% of world production.
- About 11,000-15,000 dried cloves makes 1 kg.
- Alternate bearing habit.
- Self- pollination
- Strictly tropical plant
- 2nd most important spice next to Black pepper in the world.
- Clove oil is also called ‘Caryopil oil’ and ‘amboyna
- Tree medium size, evergreen reaching upto 20 m.
- Single seeded drupe
- Antiseptic and anti-bacterial properties

7. Nutmeg

- Major producing areas is Indonesia
- Pollination by wind (mostly) and insect
- Nutmeg - dried kernel of seed have oil content from 5 to 15% of the seed weight.
- Mace - dried aril surrounding the seed have oil content from 20 to 35%
- Harvesting-June-August.
- Tree: conical reaching a height of 4 to 10 meters:
- Average weight of individual nutmeg is 60 g.
- Seed weight-6-7 g, Mac - 3-4 g
- Unproductive malé is converted into female by top working
- Pericarp-rich source of pectin (12-14%).
- Volatile oil from leaf has -Weedicidal property
- Intercropping in coconut based cropping system fits well for nutmeg
- Cinnamon, nutmeg, clove and allspice are important tree spice

8. Cinnamon

- Oldest known spice.
- Tejpat- (*Indian cassia*) - Cinnamon tamala/India tejpat
- Dried bark of C. verum is the true cinnamon
- Harvesting-Sept-November
- Bark after drying forms a quill
- Srilanka largest producer and exporter of cinnamon bark.
- Shillong, Meghalaya is the main market of Tejpat in India

9. Fenugreek

- Nipping
- Fenugreek bowl of India- Rajasthan
- Seed contains diosgenin which is used to prepare contraceptive pills.
- Diosgenin is starting material on the synthesis of sex hormone
- More than 80% area and production is contributed by Rajasthan.
- Sowing time-last week of Oct.
- Seed rate 25 kg/ha
- It is 3rd largest seed spice in India after coriander and cumin.

Two species

- Common Methi - *T. foenum-graecum*
- Keturi Methi - *T.corniculata*-orange yellow Colour
- Blue methi- *T. coerulea*
- Fenugreek is both a tropical and temperate crop.
- India is major producer and exporter of fenugreek seed.

10. Coriander

- Also called Chinese parsley
- Linalool
- Andromonoecious
- Rajasthan leads in area and production.
- Drying is done to maintain moisture level below 13% in seeds.
- Seed rate 25-30 kg/ha (Rain fed 12-15 irrigated) 40-42 kg/ha (leaf vegetables).
- Leading seed spice crop grown in India.
- Fruits - Schizocarp
- Most heavily consumed herb plant.

11. Cumin/ Safaid zeera

- Flower- white to pink
- Best sowing time-November
- Seed rate 12-15 kg/ha
- Fruits rich in thymol.
- Iran-main cumin exporter (green cumin)
- Cumin seed s are largely use as condiment
- "Witches broom" of cumin caused by phytoplasma
- Black cumin (*Nigella sativa*) carbony compound 'nigellone' found in seed

12. Fennel

- Allogamous
- Gujarat and Rajasthan being the major funnel growing states.
- Best sowing time Mid Sept-Mid Oct.
- Seed rate 10-12 kg/ha

- Bitternes- Fenchone
- Sweethers - Anethol

13. All spice/ Pimenta

- Polygamodioecious in nature
- Double Seeded Berry
- Indigenous to West Indies
- Jamaica major producing country
- Berries Vitamin-A (144SU)
- It contains flavour of cinnamon, clove and nutmeg.
- Edible part-dried immature fruits (Berries)
- Jamaican pimento is mostly used for oil extraction

14. Curry leaf

- Crystalline glucosides- Ksenigin from leaves
  -Muragin from flowers.
- Self pollinated
- Rich in Ca.

15. Kokum

- Kokum contains Hydroxyl citric acid
- Fish preservative
- 'Solkondi' is home preparation

16. Vanilla

- Vanilla is one of the most expensive spices traded globally
- Natural pollination is impossible due to presence of Rostellum
- b/c prevent contact of anther cap stigma.
- Vanillin was first isolated from vanilla beans by Gobley in 1858
- Oil is called Balsam of vanilla (vanillin) 2.3-2.9% in beans,
- Vanillin is developed due to action of enzymes. (β-glycoside).
- Vanillin is extracted by hydro alcoholic extraction.
- Coumarin - Cheap substitute for vanilla.
- Tender, green pods are harvested for use.
- Only *Vanilla planifolia* is cultivated in India

17. Saffron

- **Most expensive spice in the world.**
- In India-only cultivated in Kashmir under rain conditions.
- kennel shaped stigma is used as spice.
- Most expensive saffron shahi saffron.
- About 1.5 million flowers on drying give 1 kg of saffron
- Yield 5 kg dried saffron/ha
- Main uses cooking and colouring foods and in perfumes.
- Planting season July to September
- Harvesting season: Oct-Nov

18. Chilli

- Bird chilli- *Capsicum fruitescens*
- Variety suitable for Pickle Sindhur
- Variety suitable for extraction of capsicin - Aparna, Pachas yellow.
- Green chillies retard cancer due to natural enzyme - Asperginase

19. Asafoetida

- Stage of tapping foliage turns green to yellow
- Hadda-most priced and strongest, trade.
- Ayurvedic medicine-"Hingashakt" is prepared from asafoetida.

20. Juniper

- *Juniperous communis*
- Juni-perin mixture of tannins and sugar, is obtained from Juniper.

21. Tamarind

- *Tamarindus Indica* L.
- Tartaric acid

22. Bishop's weed/ Ajwain

- Carum seed/ carum ajowan
- Makara is used as an adulterant in ajwain seed

23. Garcinia/ Gambodge

- *Garcinia gummi-gutta*
- Clausuaceae

CHAPTER-15

Disease Management of Different Vegetable Crops

(A) Potato

Sr. No.	Common Name	Causal Organism	Scientific Name	Remarks
1	Late blight	Fungus	*Phytopthora infestans*	Most serious disease of potato Immune variety Kufri Jyoti, Kufri Kanchan Kufri Sherpa Kufri Kuber - Resistant variety **Control**- Spray 0.2% Dithane Z-78
2	Early blight	Fungus	*Alternaria solani*	Kufri Pukhraj, Kufri Sindhuri- Resistant variety
3	Black Scurf	Fungus	*Rhizoctonia solant/Pellicularia filamentosa*	Kufri Kuber - Resistant variety
4	Wart	Fungus	*Synchytrium endobioticum*	Soil borne, Endemic in Darjeeling hills
5	Brown rot of potato	Fungus	*Ps. Solancearun*	-
6	Dry rot	Fungus	*Fusarium spp.*	-
7	Potato Scab	Bacterial	*Steptomyces spp.*	Alkaline soil-Favourable
8	Bacterial wilt	Bacterial	*Ralstonia solanacerum*	-
9	Soft rot	Bacterial	*Erwinia crotovera*	-
10	Latent or Faint mosaic	PV-X and S	-	Transmitted by Aphids
11	Severe mosaic	PV-Y	-	Transmitted by Aphids
12	Leaf roll	Virus	-	Transmitted by Aphids

(B) Tomato

Sr. No.	Common Name	Causal Organism	Scientific Name	Remarks
1	Spotted wilt	Virus	-	Transmitted by thrips Most serious disease Reduced by application CCC (500 ppm)
2	Fruit rot/Buck eye rot	Fungus	*Phytopthora spp.*	Brownish spot on the fruit on the ground touching portion
3	Leaf curl	Virus	-	Transmitted by white fly
4	Mosaic	Virus	-	Transmitted by contact and seed
5	Fern leaf	Virus	-	Transmitted by Aphids

(C) Brinjal

Sr. No.	Common Name	Causal Organism	Scientific Name	Remarks
1	Little leaf	Mycoplasma	-	Transmitted by leaf hopper

(D) Chilli

Sr. No.	Common Name	Causal Organism	Scientific Name	Remarks
1	Anthracnose or dieback	Fungus	*Colletotrichum capsici*	-
2	Leaf curl	Virus	-	Transmitted by white fly thrips
3	Mosaic	Virus	-	Transmitted by Aphids

(E) Okra

Sr. No.	Common Name	Causal Organism	Scientific Name	Remarks
1	Yellow. vein mosaic	Virus	-	Most serious disease Transmitted by white fly
2	Enation leaf curl	Virus	-	Transmitted by white fly

(F) Cucurbits

Sr. No.	Common Name	Causal Organism	Scientific Name	Remarks
1	Powdery mildew	Fungus	*Erysiphe cichoracearum*	-
2	Downy mildew	Fungus	*Pseudoperanospora cubensis*	-
3	Witches broom	-	-	Transmitted by leaf hoppers Mostly occurs in bitter gourd
4	Green Mottle Mosaic	Virus	-	Transmitted by seeds Mostly occurs in cucumber
5	Bud Necrosis	Virus	-	Transmitted by seeds Mostly occurs in water melon

(G) Cole Crops

Sr. No.	Common Name	Causal Organism	Scientific Name	Remarks
1	Black rot	Bacteria	*Xanthomonas compestris* pv. *compestris*	V-shape chlorosis on margin of leaves
2	Curd rot or soft rot	Bacteria	*Erwinia crotovera*	Most destrutive during storage
3	Stalk rot	Bacteria	*Sclerotonia scleratium*	-
4	White rust or White blisters	Fungus	*Albugo candida*	Acidic soil-favourable
5	Black leg or dryrot	Fungus	*Phomalingum*	Transmitted by seed

(H) Peas

Sr. No.	Common Name	Causal Organism	Scientific Name	Remarks
1	Powdery mildew	Fungus	*Erysiphe polygoni*	-
2	Ascochyta foot rot	Fungus	*Ascochyta pisi*	-
3	Wilt	Fungus	*Fusarium oxysporum pv pisi*	-
4	Rust	Fungus	*Uromyces pisi*	-

(I) Beans

Sr. No.	Common Name	Causal Organism	Scientific Name	Remarks
1	Anthracnose	Fungus	*Colletotrichum lindemuthianum*	-
2	Web blight	Fungus	*Rhizoctonia solani*	-
3	Angular leaf spot	Bacteria	*Isariopsis griseola*	-
4	Floury leaf spot	Bacteria	*Ramularia phaseoli*	-
5	Yellow flecks	MLO's	-	Transmitted by white fly (Cowpea)
6	Phyllody	MLO's	-	Transmitted by leaf hoppers (French Bean)
7	Golden Mosaic	Virus		Transmitted by white fly (French Bean)

(J) Onion and Garlic

Sr. No.	Common Name	Causal Organism	Scientific Name	Remarks
1	Purple Blotch	Fungus	*Alternaria porii*	Most favourable temperature for disease is 28-30°C Resistant Variety Nasik Red. Arka Kalyan
2	Bottom rot or Basal rot	Fungus	*Fusarium oxysporum*	-
3	Black mould	Fungus	*Aspergillus niger*	Very common storage disease
4	Yellow dwarf	MLO	-	-

CHAPTER-16

Pest Management of Different Vegetable Crops

Sr. No.	Crop	Common Name	Scientific Name	Remark
1.	Potato	1. Tuber moth	*Pthorimea/ Ghorimoschema opercullela*	30-70% damage in tuber in storage
		2. Cutworm	*Agrotis ipsilon*	
		3. Aphids		Vector of potato virus
		4. Mite	*Hemitarsonemus*	Major pest in Deccan Plateau
		5.Cost Nematode or Golden Nematode	*Globodera spp.*	Kufri Suvarna –Resistant variety
2.	Tomato	1. Fruit borer	*Helicoverpa armigera*	
		2. White fly	*Bemisia tabacii*	transmitted by leaf curl virus
		3. Serpentine leaf minor	*Liriomyza trifoli*	-
3.	Raddish	1. Mustard saw fly	*Athalia lunnus*	
		2. Pointed bug.	*Bagrada cruciferum*	-
4.	Okra	1. Shoot and Fruit borer	*Earias vittela*	-
		2. Leaf hoppers	*Amarasca biguttula*	Most serious pest of okra
		3. Blister Beetle	*Mylabris pustulata*	-
		4. Stem and Bulb Nematode	*Ditylenchus dinsaci*	-
5.	Garlic & Onion	1. Thrips	*Thrips tabaci*	-
		2. Mite	*Aceria tulipae*	-
		3. Onion fly	*Delia antiqua*	-
		4. Bulb nematode	-	-
6.	French Beans & cowpea	1. Stem fly	*Ophiomyia phaseoli*	80-90% mortality of plant

Sr. No.	Crop	Common Name	Scientific Name	Remark
7.	Cucurbits	1. Red Pumpkin Beetle	*Aulacophora/ Raphidopera foveicollis,*	-
		2. Fruit fly	*Bactocera cucurbitae*	Arka Suryamukhi-Resistant variety
		3. Thrips	*Thrips tabaci*	-
8.	Chilli	1. Thrips	*Scirtothrips dorsalis*	Transmits leaf curl of chilli
9.	Colecrops	1. Dimmond Black moth	*Plutella xylostella*	Most damaging pest
		2. Stem borer	*Hellula undalis*	-
10.	Brinjal	1. Fruit and shoot borer	*Leucinodes orbonalis*	-
		2. Stem borer	*Euzophera perticella*	-
		3. Ash weevil	*Myllocerous subfaseiutus*	Saw like damage in the margin of leaves
		4. Epilachna (Hadda) beetle	*E. vigintictopunctata*	Tolerant variety-Punjab Barsat
11.	Leafy vegetable	1. Leaf eating caterpillar	Hymenia recurvalis	-
12.	Tomato & Brinjal	1. Root knot Nematode	Melidogyne spp.	Tomato Resistant Variety-Hissar lalit, Arka vardan, Mangla Selection-120 Brinjal Resistant Variety Black beauty, Mangrigota
13.	Sweet Potato	1. Sweet potato weevil	*Cylas formicarius*	Monophagous pest
14.	Black pepper	1. Pollu beetle	*Longitarsus nigripenis*	40% damage in Back Pepper
15.	Cardamon	1. Aphids	*Pentalomia nigronervroa*	Transmit vector of katte disease
		2.Shoot and capsule borer	*Dichocracis puntiferalis*	-
16.	Turmeric & Ginger	Rhizome-Fly	*Mimegralla coeruleferons*	Cause Rhizome rot of Ginger

CHAPTER-17

Physiological Disorders of Different Vegetable Crops

Sr. No.	Crop	Name of Disorder	Cause
1	Potato	1. Internal Brown spot	Moisture deficiency
		2. Greening	Excessive exposure to sunrays
		3. Black Heart	Poor Ventilation/O_2 deficiency
		4. Hollow Heart	Excessive Nitrogen
		5. Chilling injury	Low temperature 0°C
		6. Freezing injury	Low temperature (–1 to-2°C)
2	Tomato	1. Cracking	Boron Deficiency (Spray Borax .25%)
		2. Blotchy ripening	Potassium deficiency
		3. Blossom end rot	Ca deficiency+High temperature+ irregular moisture supply
		4. Sun scald	Excessive exposure to sun-rays
		5. Puffiness or pocket	low or high temperature + lack of fertilization
		6. Cat face	Abnormal growing conditions
		7. Golden flake	excess of Ca-oxalate/low K: Ca ratio
		8. Radical cracking	
		9.Crease stem	Boron Deficiency
3	Cauliflower	1. Ricyness	Fluctuation in temperature, High Humidity, Excess Nitrogen
		2. Fuzziness	Cultivation in abnormal time.
		3. Blindness	Frost
		4. Leafyness	High temperature
		5. Buttoning	Nitrogen deficiency, sowing early variety in late season
		6. Hollow stem	Excessive Nitrogen
		7. Whiptail	Molybdenum deficiency, common in Acid soil
		8. Chlorosis	Magnesium deficiency
		9. Brown rot/red rot	Boron deficiency

Sr. No.	Crop	Name of Disorder	Cause
4	Carrot	1. Splitting	Boron deficiency. Excessive Nitrongen, Change in soil moisture
		2. Cavity spot	Calcium deficiency, Excessive Nitrogen, change in soil moisture
		3. Bitterness	Excessive ethylene
		4. Forking/ Elangated root	Hard soil pan, undecaying manure
		5. Pithiness/Pore extend	Delay harvesting
5	Beet root Radish	1. Brown Heart or crown heart or heart rot	Boron deficiency
6	Water Melon	1. Blossom end rot	Ca deficiency+High temperature+irregular moisture supply
7	Colocasia	1. Metsubre	Ca deficiency
8	Sweet potato	1. Growth crack	Moisture Imbalance
9	Celery	1. Black Heart	Calcium deficiency
		2. Cracked stem	Boron deficiency
		3. Pencil strip	Excess of Phosphorus
10	French Bean	1. Blossom drop	High temperature
		2. Hypocotyl cracking (necrosis)	Calcium deficiency
11	Garlic	1. Bulb sprouting	Excess Nitrogen + soil moisture
		2. Splitting	Delays Harvesting
12	Lettuce	1. Tip burn	Ca deficiency + Unfavourable climate
		2. Rossette spotting	Ethylene injury
13	Chilli	1. Blossom end rot	Excess Nitrogen + water stress
		2. Frog eye rot	-
14	Raddish	1. Akashin	Boron deficiency
15	Cucumber	Pillow	Calcium deficiency.
16	Brijal & Okra	No disorder	-

CHAPTER-18

Storage Temperature and Relative Humidity for Different Vegetables

Vegetables	Temperature (°C)	Relative humidity (%)	Storage life (weeks)
Tomato (un- ripe)	8.9-10	85-90	4-5
Tomato (ripe)	7.2	90	1.0
Brinjal	10-11.1	92	2-3
Chilli (green)	7.2	58-90	3-5
Chilli (ripe)	7.0-7.2	90-95	2.0
Potato	3.0-4.5	85	34
Cabbage (early)	.0.0-1.7	92-95	4-6
Cabbage (late)	0.0-1.7	92-95	12
Cauliflower (late)	0.0-1.7	85-95	7.0
Bitter gourd	0.6-1.7	85-95	4.0
Cucumber	10.0-11.7	92	2.0
Muskmelon	1.7-3.3	85-90	1.5
Pumpkin	10.0-12.0	70-75	24-36
Winter squash	12.8-15.6	70-75	24-36
Watermelon	7.2-15.6	80-90	2.0
Beet (topped)	0.0 -1.7	90-95	8-14
Beet (bunched)	0.0	90	1.5
Carrot (topped)	0.0	95	20-24
Radish (topped)	0.0	90-92	3-5
Turnip	0.0	90-92	8-17
Lima bean	4.5-7.2	90-95	2
Pea	0.0	88- 92	2-3
Garlic	0.0	65	28-36
Onion (bulb)	0.0	70-75	20-24

Vegetables	Temperature (°C)	Relative humidity (%)	Storage life (weeks)
Onion (leaves)	0.0	90-95	2.0
Taro	11.2-12.5	85-90	22
Cassava	0.0-1.7	85	23
Sweet potato	10.0-12.8	80-90	13-20
Okra	8.9	90	2

Post Harvest Management of Vegetable Crops

1. Potato

- Suberization is the process in which wounds are healed. A temperature of 25°C with 95% RH are ideal for suberization.
- Kufri Chandramukhi and kufri Bahar have short dormancy.
- Kufri Lalima and Kufri Sindhuri have long dormancy.
- MH (desprout) is commonly used sprout suppressent in India.
- Tuber moth causes maximum damage during storage when stored under high temperature.
- High humidity and condensation of water on tuber surface can lead to infection of soft rot causing bacteria, (*Erwinia spp.*)
- Cold storage can accommodate only about 45% of the total potato production.
- Heaps and pits are most common methods of potato storage in plains and plateau region.

2. Cassava

- Starch is the major industrial product from cassava.
- Thailand and Indonesia are the world's largest exporters of Cassava pellets.
- Cassava tuber contain cyanoglucosides (linamarin and lotaustralin) which are converted to toxic compounds hydrocyanic by the endogenous enzyme linamarase

3. Sweet potato

- It contains tripsin inhibitor which inhibits the important digestive enzyme trypsin.
- Corms of Elephant foot yam contain oxalic acid.
- Starch of arrow root is used mainly as an infant food.

CHAPTER-19

About Seed Technology and Breeding

Seed

Botanically seed is a fertilized ovule consisting of embryo, endosperm and seed coat and have germination capacity.

- National seed act-1966 and force 2nd October 1969, now Indian seed act, 2004 and force Jan., 2005.
- Seed act, 2004 includes Agriculture, horticulture, forestry, plantation, medicinal and aromatic plants.
- The seed act, 2004 permits sale of certified seeds.
- **As per** seeds act, 2004 the variety of field crops can be registered for 15 years.
- The variety of perennial crops (Tree and vines) can be registered for 18 years

National seed corporation (NSC): established 1961

- It was registered 1963 having objective of –
- To promote development of a seed industry of India
- To promote and supply foundation seeds of various crops

State seed corporation (SSC): it concern with

- Production and supply of certified seed
- To reduce the work load of NSC

Classes of Seed

Particulate	Nucleus	Breeder	Foundation (Mother Seed)	Registered	Certified
Produce by	Plant breeder	Plant breeder	NSC	SSC	SSCA
Grown in India				×	
Progeny	Initial categories of improved seed	Nucleus	Breeder	Foundation	Both Foundation and Registered
Seed certification	×	×			
Genetic purity	100%	100%	100%	100%	100%
Physical purity	100%	100%	98%	98%	98%
Tag colour	×	Golden/ Yellow	White	Purple	Azure blue
Size	×	12 × 6	15 × 7.5	-	15 × 7.5
Commercial cultivation	×	×	×	×	

- **NSC:** National Seed Corporation
- **SSC:** State Seed Corporation
- **SSCA:** State seed certification agency
- **NSP:** National seed project, New Delhi,1977
- **SVRC:** State variety release committee
- **CVRS:** Central variety release committee
- **ISTA:** International seed testing association
- **FAO:** Food and agriculture organization
- **UPOV:** Union for the protection of new plant varieties, Geneva,1978
- **PBR:** Plant breeder right
- **SFCI:** State form corporation of india
- **IMSCS:** India minimum seed certification standard,1988
- **AOSCA:** Association of official seed certification agency
- **MES:** Main experimental station
- **SIF:** Student Instruction form

- **NSPF:** National Seed project form
- **CRS:** Crop research station
- **NSRTC:** National seed research and training system
- **IISS:** Indian research of seed science, Varanasi
  - ☆ **Father of Seed Technology :** Rabb Fraley
  - ☆ **Father of Seed Testing:** F.Nobbe
  - ☆ **Father of Seed Pathology:** Poul Neergard
    - **PPV&FR Act, 2001**, it approved in 2004 and come into force 2005
    - **HQ**- New Delhi
    - **Obj.**- To encourage development of new plant varieties and promote the development of seed industry in India.
  - ☆ **Seed quality index**- Vigour of seed germination
  - ☆ **Real value of seed**- Purity % x Germination% /100
  - ☆ **Impurity %**=Dockage.
  - ☆ **Genetic purity-** Grow out test (GOT)
  - ☆ **Seed viability test**- Tetrazolium test (Biological test)
    - Dehydogenase enzyme
    - Bright red colour
    - Colour become more intense of embryo
    - Quickest method
    - **Godex test**- Seed germination indicator

Type of Seed

- **Orthodox-** <5% Moisture content.

  Most of agricultural crops
- **Semi recalcitrant**- Not <25% Moisture content.
- **Recalcitrant**- Not <30% Moisture content.

  Most of horticultural crops

Breaking of dormancy

- **Scarification-** KNO_3 (1-3%) strongest dormancy breaker
- **Stratification-** Chilling Temp. <7°C

- **Purity%=** $\frac{\textit{Wt. of pure seed}}{\textit{Total wt. of working sample}} \times 100$

Crop	Purity (%)
Okra and Cucurbits	99%
Tomato, Cauliflower, Onion, Chilli, Sweet peper, Radish, Brinjal	98%
Carrot	95%

- **Germination %=** $\frac{\textit{Total number of seed germinated}}{\textit{Total number of seed plated}} \times 100$

Cowpea	75%
Tomato	70%
Cauliflower and Okra	65%
Cucurbits, Chilli, Carrot	60%

- **Mode of reproduction**
  - The process by which all living organism give rise to offspring of similar kind is called reproduction
  - Asexual- without fusion of male and female gamete
  - Sexual-fusion with male & female gamete

(A) Asexual Reproduction

Vegetative- multiplication of plant by various plants Part like stem, root & leaves

Modified underground stem is of four types.

- **Rhizome**
  - Turmeric (*Curcumus langa)*
  - Ginger (*Ginger officinalis*)
- **Corm**
  - Arvi - Eddoe type -*C. esculenta var. antiquorum*
  - Dasheen type- *C. esculenta var. globulifera*
- **Tuber**
  - Potato - *Solanum tuberosum*
- **Bulb**
  - Onion - *Allium cepa*

- Garlic- *A. sativum*

Apomixis – Term- Hacke 1893

- Discover by -Winkler 1906
- It is part of asexual reproduction.
- Development of seed without fertilization is called Apomixes
- If Apomixis occurs alone in plant is obligate apomixis'
- It occurs sexual reproduction is facultative apomixis'

Apomixis is of four type

- **Parthenogenesis-** Term was given by Owen (1849)

  The development of embryo from Egg Cell without fertilization.

  Parthenogenesis is found or Drone bees and Aphid
- **Apogamy-** Tem was given by winkler In 1908

  The development of embayo either from Synegid cell or Antipodal cell without Fertilization.
- **Apospory-** Tem was given by Druery & Brower 1886

  The development of embryo Sue from Any somatic cell without meiosis
- **Adventive embryony-** Tem was given by Strusberger, 1878

  The development embryo from any somatic cell of ovule without embryo sac formation

(B) Sexual Reproduction

- Sexual reproduction is known as Amphimixis.
- The term "Amphimixis' was coined by Weisman(1891)
- When embryo is developed by the fusion of male & female gamete is called sexual reproduction.
- It is of two types,viz., autogamy and allogamy
- **Autogamy-** Transfer of pollen grains from the anther to the stigma of the same flower, also called self pollination.

Mechanisms which promote autogamy:

- Bisexuality
- Homogamy
- Cleistogamy *eg.* Lettuce, peas and bean
- Chasmogamy *eg.* Tomato

Allogamy-Transfer of pollen grains from the anther of one plant to stigma of another plant.

Mechanisms which promote allogamy

- Separation of sex *eg.* Cucurbits
- Herkogamy- *eg* Lima beam
- Dioecy- Pointegourd , Asparagous, spinach
- Dichogamy
- Heterostyly *eg.* Brinjal and Primula
- Male sterility
- Self incampatibility

Important point

Pureline- progeny of a single self pollinated homozygous plant obtain by Selfing.

- It is used to produce Hybrid.
- Example of homozygous and homogenous

Inbred line- In Allogamous crops, true breeding (homozygous Line) obtained by continuous selfing.

- It is used to produce hybrid seed.
- Example of homozygous and homogenous

Clone - progeny of single plant obtained by asexually propagation.

- It is a example of heterozygous & homogenous

Cybrid - Somatic hybrid with normal protoplast of one species and nucleusless protoplast of other species.

Concept **Cybrid** given by Bunn *et al.*,1974

Vybrid- progenies obtained from crossing two facultative Apomitic

Diverse cytoplasm can be fixed by Protoplasm fusion

Top Crass-

- It was given by Device in 1927.
- Crossing of an Inbred with OPV. (Open pollinated variety)
- Also known as Inbred variety cross.
- It is used to test the "GCA" of any Inbred.

Double Top Crass-

- Crossing of a single crass with OPV
- DTC= (AXB) X OPV

Hybrid- progeny of a cross b/w genetically dissimilar genotype

- When F_1 used for commercial cultivation is known as hybrid variety.

It is of several type

Single cross hybrid-

- Crossing b/w two different Inbred.
- In Single cross hybrid involve only two Inbred.
- **Maximum uniformity is found.**
- It is used to test the SCA of any Inbred.
- Single crass A x B
- All possible single crass con be calculated by n(n-1)/2

Double Crass hybrid

- Crossing b/w two single crass
- In case of double cross hybrid involve four Inbred
- All possible double cross can be calculated by n(n-1) (n-2) (n-3)/8

Triple crass hybrid-

Crossing be single cross with an Inbred

Involve only three Inbred.

All possible triple cross can be calculated by n(n-1)(n-2)/2

Back Crass- Crossing of F_1 with either of its parent.

- Isogonics lines are developed through Back cross
- When F_1 crossed with homozygous recessive parent, it is known as test cross.
- Test Cross term given by (Bridge)
- Back cross method is applicable for three groups of crop plants, *viz.,* self pollinated, cross pollinated and asexually propagated.
- It is widely used for development of **disease resistant varieties.**
- It also used been used for inter-specific gene transfer and development of multiline varieties in self pollinated varieties
- **Hybridization**
    - ☆ Intra specific/Intervarital – crossing b/w two different variety
        - Fully fertile
    - ☆ Interspacific/Interagenic- crossing b/w two different species of the same genous
        - Partial fertile

- ☆ Intergenric- crossing b/w two different genus of the same family
  - Sterile
- ❒ **Emasculation-** removal of immature anther from a bisexual flower.

Method of emasculation

- ❒ Hand emasculation
- ❒ Emasculation with cold water and hot water
- ❒ Alcohal
- ❒ Suction method
- ❒ Chemical
- ❒ Male sterility
- ❒ **Self incompatibility-** Inability of a plant with functional pollen to set seed when self-pollinated.
  - ☆ The term SI was coined by Stout, 1917
  - ☆ First reported in *Verbascum phoenicium* by Koelreuter
  - ☆ Floral morphology (i) Heteromorphic- eg – Primula

    (ii) Homomarhpic - due to physiological or genetic

Gametophytic S.I.

- ❒ Self incompatibility which is controlled by the genetic constitution of gametes.
- ❒ GSI first discovered by East and Mangelsdorf (1925) in *Nicotiana sanderae*
- ❒ It is found in tomato and sugarbeet.
  - ☆ Fully incompatible (S_1 S_2 x S_1 S_2)
  - ☆ **Partially incompatible (S_1 S_2 x S_1 S_3)**
  - ☆ Fully compatible (S_1 S_2 x S_3 S_4)

Sporophytic S.I.

- ❒ Self incompatibility which is governed by the genotype of pollen producing plant, *i.e,* sporophyte.
- ❒ SSI first discovered by Hugenes and Babcock in 1950
- ❒ It is not successful in Chinese cabbage
  - ☆ Fully incompatible (S_1 S_2 x S_1 S_2)
  - ☆ Fully compatible (S_1 S_2 x S_3 S_4)
  - ☆ Compatible**(S_1 S_2 x S_1 S_3)**

Distyly- Style having two positions, *viz.,* low, and high.

Tristyly- Style having three positions, *viz.,* low, medium and high.

Monoallelic SI - controlled by single gene.

It is found in family -

- Leguminosae,
- Solanaceae
- Cruciferae.

Diallelic SI- SI controlled by two genes.

It is found in the family Graminac.

Polyallelic SI- SI governed by >21 genes.

It is found in **Sugarbeet**.

- **Male sterility**- A condition in which either pollen is absent or non-functional in flowering plants.
  - ☆ First reported by Koelreuter (1963) in flowering plants
  - ☆ MS is controlled by single recessive gene.
- It promote allogamy and prevent autogamy.
- MS is cause by nuclear gene,chloroplast gene and mitochondrial gene
- MS is used for development of hybrid seed
- In case MS only three type line are used
- **A line-** The male sterile line.
- **B line-** Isogenic line of A line with male fertility.
- **R line-** A line which restores fertility when crossed with cytoplasmic genic male sterile line. hermaphrodite individuals in
- Genetic male sterility (GMS)- Pollen sterility which is caused by nuclear genes.
- Cytoplasmic male sterility (CMS)-Pollen sterility which is caused by cytoplasmic genes.
- Cytoplasmic genic male sterility (CGMS) -Pollen sterility which is controlled by both cytoplasmic and nuclear genes.
- **Heterosis-** The superiority of F_1 in one or more characters (especially yield) over its parents.
- It is also called hybrid vigour
- The term heterosis was coined by Shull (1914)
- It is was two type- positive and negative **heterosis**

A/c to power concept 1944

Parent and their performance unit	Performance of F_1 unit	Remark
Parent A (10)	>10	Heterosis
	10	Complete dominance
	10 to 8	In complete dominance
Mid-Parent (8)	8	No dominance
	8 to6	In complete
	6	Complete dominance
Parent B (6)	<6	Heterosis

❐ **Some term related to Heterosis**

Hetrosis result due to (i) Dominance
(ii) Over dominance
(iii) Epistasis

Hetrosis can be fixed by (i) asexually reproduction
(ii) Apomixis
(iii) Polyploid
(iv)Balanced leathal system

Luxuriance- The superiority of F_1 over its parents in vegetative growth, but not in yield and adaptation

❐ It is also called pseudo-heterosis.

Mid-parent heterosis- Heterosis over the mean value of both the parents.

❐ **It is** also called mean heterosis.

Heterobeltiosis- Superority of F_1 hybrid over the better parent for a particular character.

Economic heterosis- Superiority of F_1 over the popular variety of a region.

❐ It is also called useful heterosis.

Dominance Hypothesis- Heterosis due to superiority of dominant alleles over recessive alleles.

❐ Heterosis is directly proportional to the number of dominant genes contributed by each parent.

❐ **Overdominance hypothesis**- Heterosis due to superiority of heterozygote over both the homozygotes.

❐ **Epstasis-** Interaction between alleles of two or more different loci.

❐ **Inbreeding**- Mating between closely related individuals such as selfing and sib- mating

- **Inbreeding depression-** Loss or decrease in vigour and fitness as a result of inbreeding Hybrid.
- The progeny of a cross between genetically dissimilar parents or genotypes.
- **Hybrid variety-** The F_1 population which is used for commercial cultivation.
- Hybrid varieties are of two types, *viz,* intraspecific hybrid and interspecific hybrid.

Hetroploidy- Any change in the chromosome number from diploid state.

- ☆ **Aneuploidy-** gain or loss of one or few chromosome of the genome

		Single	Double	Gamet
Hypoploidy (loss)	**Nullisomy**	**2n-2**	2n-2-2	n-1
	Monosomy	2n-1	2n-1-1	n-1, n
Hyperploidy (gain)	Trisomy	2n+1	2n+1+1	n+1,n
	Tetrasomy	**2n+2**	2n+2+2	n+1

The term Recurrent selection (RS) was given by Hull,1945

Phenotype RS	Tester	Natural tester	Scientist
Simple RS	×	homozygous	-
RS for GCA		1 hetrozygous(OPV)	Jenkins,1940
		1 homozygous (Inbred)	Hull,1945
		2 hetrozygous	Comstock, Robinson and Harvey,1949

- **Inbreeding:** Also known as selfing/back crossing/sib-mating
  - ☆ Loss in vigour and fitness due to selfing
  - ☆ Ofter selfing homozygosity is increased
  - ☆ %ID= F1-F2/100×100
  - ☆ High ID in Carrot and Onion
  - ☆ Low ID in All cucurbits
  - ☆ Lack of ID in all autogamous crops

CHAPTER-20

Protected Cultivation of Vegetable Crops

- India area under green houses cultivation of horticultural crops around not >10000 ha
- High value and low volume vegetable are grown
- China, Japan, Spain, Italy has the highest area under greenhouse
- In india MH and KN has the highest area under greenhouse
- Two main branched pruning system in green house: Sweet pepper
- Glass houses very popular in Netherlands.
- CPCT - Centre for protected cultivation is located at IARI, New Delhi.
- Polyethylene film widely used in green houses
- Polyethylene is highest light transmission % of glazing materials
- Low cost or naturally ventilated green houses- Suitable for **cucumber**
- Walk in tunnels- Low cost semi- permanent protected structures eg. cucurbits
- Insect net proof houses used for virus free crop production- Tomato and sweet pepper
- Rain shelters- Protection tomato plants against the impact of heavy rainfall
- Most popular type of green houses Gutter connected structure
- Reduction of green house temperature by using evaporative cooling method
- Optimum CO_2 for crop high yield 1000-1500 ppm

Net houses

- Shade nets- Prepared from HDPF (High density polyethylene) shade intensity varies between (25-75%)
- Insect proof nets 40-60 mesh Free from viruses)

Plastic mulches

- Coloured plastic mulch is reduce the insect damage
- Yellow mulch is repels the whitefly
- Black plastic mulch is used for soil moisture conservation and weed control
- Transparent mulch 30-40 micron thickness) is increase the soil temperature (Soil solarization)

High tech nursery for vegetable crops

- Suitable combination media for root medium Coco Peat, vermiculite and perlite (3:1:1)
- Rooting media for plug tray nursery - Coco-peat: Good drainage and porosity

  -Vermiculite Chemically hydrated magnesium aluminum silicate
- Rich in Ca, Mg

  -Perlite –Neutral

Vegetable grafting

- Grafted vegetables popular in Japan and Korea
- Leading grafted vegetable in the world- water melon
- Recommended level of light for grafting acclimatization :3-5 lux
- Emergeneric grafting popular in cucurbits

Cucurbits

- Most widely used cucurbit rootstock Shintozwa- Suitable for cucumber, melon and water melon
- Shintozwa is derived from (*C. maxima x C. moschata)*- Resistant to fusarium wilt

Brinjal

- Interspecific grafting popular in brinjal
- Commercial brinjal rootstock *Solanum integrifolium* (Resistant to Verticillium and Fusarium wilt)
- **Brimato:** The Brinjal hybrid- Kashi Sandesh and improved cultivar of tomato-Kashi Aman were successfully grafted into brinjal rootstock- IC111056
- **Pomato:** (Potato+Tomato)

Hydroponics in Vegetable Crops

- The term hydroponics was derived from two Greek words: Hydro-water and pones-labour

- Hydroponics- The system of growing plants in a nutrient solution
- Father of hydroponics F. Gericke, University of California, USA
- Hydroponics was first developed on commercial scale by W.F.Greicke, University of California, USA (1930)
- Hydroponics communalized in 1960-70's
- India first introduced at Kalimpong, Darjeeling District, West Bengal
- Hydroponics has 3 methods: Water culture, Drip culture and Gravel culture
- Drip culture is also known as sand culture
- **Soil less culture-** Growing plants in the media other than soil
- **Soilless culture techniques -** Nutrient Film Technique (NFT) Tube culture and (Aeroponics)
- Maximum area under hydroponics country: Israel
- First crop raised in hydroponics- Tomato
- Optimum pH of solution **(5.2-6.5)**
- Nutrient solution replaced once in 4-6 days interval
- In USA only Styrofoam flats are often used compared to plastic pro-trays

Soilless mediums

- Peat (acidic in nature) widely used medium in soilless culture
- Perlite-Grey white silicaceous material (volcanic origin)
- Vermiculite- Micaceous mineral
- Rock wool aeration to the roots
- Sphagnum moss or peat moss (Bryophytes)- High water Holding capacity

Nutrient film technique (NFT)

- True hydroponics system
- The vegetable crops are adaptable to NFT are tomato and cucumber

Aeroponics

- The plants are grown in trough or container; the roots are sus aded and spraved with nutrient mist
- Aeroponics: suitable for short stature vegetables e. Lettuce, spinach
- **Aquaponic**- is a combined structure of growing of fish and Plants

CHAPTER-21

Miscellaneous

Mineral Nutrition in Vegetable Crops

- Essential nutrient (17)
- Basic nutrients: C, H, O
- Macro nutrients: N, P, K, Ca, Mg, S
- Primary nutrients: N. P. K
- Secondary nutrients: Ca, Mg. S
- Micro nutrients: Fe, Zn, Cu, B, Mo, Cl, Ni
- Ni 17th essential nutrients in plants (1987)

Morphological markers

- Potato leaf- Tomato
- Non-lobed leaf- Watermelon
- Purple stem- Cabbage
- Brown seed coat-Onion

Sex forms

- Cleistogamy - Lettuce
- Chasmogamy - Tomato, Capsicum.
- Protogyny - Cole crops, Okra.
- Protandry - Onion, Carrot.
- Herkogamy - Lima Bean
- Spinach is tetramorphic (5 sex form is present).
- Peas and beans are self pollinated crop due to presence of keel

- **Some IARI released varieties:**
  - Tomato- Pusa Rohini
  - Sponge gourd- Pusa Sneha
  - Ash gourd- Pusa Ujjwal, Pusa Shakti
  - Cucumber- Pusa Uday
  - Snapmelon- Pusa Shandar
  - Cauliflower- Pusa Meghana
  - Carrot (Black)- Pusa Ashita
  - Leafy mustard- Pusa Saag No. 1
  - Bottle Gourd- Pusa Sandesh
  - Garden Pea- Pusa Pragati
  - Broad been- Pusa Sumeet Pusa Udit
  - Chenopodium- Pusa Bathua
- Iodine rich vegetable- Okra, Onion.
- Tapioca or cassava- highest carbohydrate present.
- C-enamal-Sulphur resistant cane (vegetable)
- B.V.O: Brominated vegetable oil
- 1st public sector commercial hybrid in vegetable - Pusa Meghdoot and Pusa Manjari in Bottle gourd in 1971 (IARI, New Delhi).
- 1st private sector commercial hybrid of tomato- Karnataka-F_1 by IAHS, Bangalore in 1973.
- 1st commercial hybrid of sweet pepper - Bharat (Public sector) in 1973.
- Agathi- high Ca.
- Recalcitrant seeds - Pointed gourd, Chow-chow
- Sweet potato is cheapest source of colorie.
- Cassava multiple hybrid variety-Sree Shaya.
- Spineless variety of okra - Pusa Sawani.
- Ash gourd has longest storage life among cucurbits.
- Home or Kitchen garden is most ancient type of vegetable garden.
- Market garden is very intensive type of vegetable garden.
- Truck garden is very extensive type of vegetable garden.

- Green leafy vegetables are rich source of folic acid.
- Vegetable requirement/capita/day - 300g (125g leafy vegetables, 100g root and tuber and 75g others).
- Golden tomato was developed by AVRDC and it is having 3 to 6 time greater beta-carotene than red tomatoes.
- French bean new variety-Arka Anoop photoinsensitive, bushy, resistant to both rust and bacterial blight.
- Pea new variety- Arka Karthik do not require staking and resistant to powdery mildew and rust.
- OrNamental peppers are often called as Chirstmas Peppers
- **Resistant sources for chill** :-
  - ☆ Thrips- NP-46A.
  - ☆ Leaf Curl- Pant C-1
  - ☆ Mite- Punjab Lal
  - ☆ Aphid- Kalyanpur Red
- In okra 4-5 ridges is desirable for breeding objective point of view.
- Rachna is powdery mildew resistant cultivar of pea.
- Powdery mildew in pea is controlled by a single recessive gene (er)
- Fern leaf disorder in potato - Zn- deficiency.
- CFTRI is situated- Mysore.
- Thickness of polythene use for mulching in vegetable cultivation:200μ
- In Brinjal ratio of short styled flower to long/medium styled flower is 3-4:1
- Club root is prevalent in acid soil (pH<7.0) Acidic soil
- Black rot (*Xanthomonas campestris*) resistant variety :-
  - ☆ Cauliflower- Pant Shubhra, Pusa Snowball- k-1
  - ☆ Cabbage- Pusa Mukta, Majabi.
- Black leg resistant variety of cabbage-Pusa Drum Head
- Pillow- (disorder of processing cucumber) due to low calcium level in the tissue.
- Leaf silvering disorder of summer squash due to moisture scarcity
- Perennial cucurbits- Chow-chow, Ivy gourd.
- King of gourd- Pointed gourd.

- Dioecious vegetables - Pointed gourd, Asparagus, Yam, Spinach, Spine gourd.
- **Enzymatic browning**
  - ☆ Polyphenol oxidase
  - ☆ Phenolase
  - ☆ Pectic methyl estarase
- Mi - gene for root knot nematode
- "Gajendra" is an improved cultivar: Elephant foot yam
- White brinjal cultivar: KKM-1
- Maturity indices for Elephant foot yam: Yellowing and dropping leaves
- Tm-gene for tobacco mosaic virus resistant in tomato (single dominant gene)
- Temperate gynoecious lines of cucumber by IARI - SR- 551F, W-12757.
- Tropical gynoecious lines of cucumber by IARI - 304-C, 319, 12, 338.
- Stability in gynoecious sex expression affected beyond 30°C in cucumber.
- Glucosinolate" associated with resistant to Downey mildew in cauliflower.
- Phenols associated with resistant to tomato leaf curl virus in tomato.
- Vegetable known as *'Kallikrein inhibitor*- Potato.
- Parthenocarpic lines in tomato- Severianin, PSET-1, RP- 75/59.
- The 1st GM (genetically modified) potato was developed at CPRL, Simla for increasing the protein content in tuber which consists of genes from grain amaranthus.
- Tendrils are absent in cucurbit-Summer squash
- Metaxenia found in Cucumber, Bottle gourd.
- Triple cross hybrid of cucumber TSkhA-2693.
- Dania a cauliflower variety selected at Kalimpong by inbreeding- Pura Deepali
- Colchicine- $C_{12}H_{25}O_6N$.
- Capsaicin- $C_{18}H_{27}O_3N$.
- Piperine- $C_{17}H_{19}O_3N$.
- Determination of TSS: By Hand refractometer
- Estimation of sugar: By Shaffer micro method
- Measuring the salt concentration: By Salometer
- Cirtic acid: Nature of acidulant
- In summer squash F_2 seeds can be used.

- Coccinia is belong to the diecious and parthenocarpy
- Tomato cultivar suited for Rainfed farming- Arka Meghali
- Marsh spot in pea due to deficiency- Mn
- Enation leaf curl is disease of -Okra
- Multivitamin green vegetable- Chekurmanis
- Transverse cotyledon cracking is an physiological disorder: French bean
- In carrot mainly 3 types of male sterility is found-
  - ☆ Brown anther type (ba)
  - ☆ Petaloid Type (pr)
  - ☆ gum' type (gu) in *Dacus carota var gummifera.*
- In carrot male sterility was 1st reported by Welche and Grimble.
- In Onion 1st male sterility was discovered by Dr. H. A. Jones from the cultivar Italian Red (15-33) and the male sterility was CMS-S type.
- A new source of male sterility in Onion was discovered from a species: *Allium galanthum*
- For hybrid seed production female male ratio

Carrot- 4:2

Onion- 4:1

- Cracking in root crops is governed by single dominant gene.
- Black rot of Carrot is caused by *Alternaria radicina.*
- In radish seeds weighting < 5 mg should not be used for steckling production.
- Japanese wild radish *Raphanus raphanistrum.*
- The vector of turnip yellow mosaic virus is flea beetle and its alternate host is cabbage.
- Nieuwhoff developed male sterility in cauliflower in 1961.
- Rundfeldt developed male sterility in cabbage in 1960.
- Strongest self-incompatibility in cole crops - Kale.
- Weakest self- incompatibility in cole crops summer cauliflower.
- Or mutant gene is responsible for accumulating Beta- carotene in cauliflower.
- Protein responsible for self incompatibility in cole crops glycoprotein.
- New world crop-cassava
- **Clipping** is done in - Amaranthus.
- Nipping is done in - Fenugreek.

- Pusa Deepali cultivar of cauliflower is self blanched.
- Mr. Onion is Dr. H.A. Jones.
- The development of double cross hybrids using S.I. System in cabbage was done by Odland and Noll in 1950.
- 1st F_1-hybrid of Brussels sprouts using S.I. system - Avon Cross.
- True breeding monoecious lines in spinach X^mX^m
- Vegetable meat is known to - cowpea.
- Boneless meat is known to- Soybean.
- Blue lake is a variety of French bean.
- International year for fruit and vegetable :2021
- Potato are borne on :Stolan
- PAU developed two new varieties of potato:Punjab Potato 101, 102
- Plant rectangularity is :Plant to plant distance or row to row distance
- Kashmir Beauty, Golden Heart are the varieties of carrot.
- Side harvesting or partial harvesting is practiced in Colocasia.
- Double harvesting is done in yam.
- Phenolic factor present in onion which has antifungal properties is - catechol.
- Among cucurbits richest source of Fe is water melon (7.9mg/100g).
- Chilli hybrids from PAU, Ludhiana using GMS system-
  - ☆ CH-1(MS-12 x LLS)
  - ☆ CH-3 (MS-12 x 2530)
- 1st hybrid of muskmelon- Punjab Hybrid - 1
- 1st hybrid of water melon - Arka Jyoti
- 1st hybrid of cucumber - Pusa Sanyog (Japanese Long Green x Green Long of Naples).
- Red Cherry variety of tomato can tolerate Chromosomal imbalance.
- In Cole crops Black leg occurs in saline soil.
- Club root occurs in acid soil.
- Hot weather hybrids of cabbage (30-35°C) - Green Boy, Green Express.
- Slicing cucumber- 20-25 cm long.
- Pickling cucumber- 8-10 cm long.
- Potato scab occurs in alkaline soil.
- River bed cultivation variety of muskmelon - Pusa Sharbati.

- Garlic variety for dehydration –(G-189)
- Parthenocarpic gene of tomato-rot-B gene.
- Parthenocarpic gene of brinjal - iaah gene DefH-9 gene
- Tropical perennial vegetable crops -Drumstick, Agathi, Curry leaf, Chekurmanis.
- Temperate perennial vegetable crops - Asparagus, Rhubarb, Sea kale, Globe Artichoke.
- Paraheliotrophy found in cowpea.
- 1 I.U. (international unit) vitamin A = 0.6 µg B-carotene.
- Tomato fruit colour is determined by colorimeter
- Cauliflower variety Pusa Deepali is early type and self blanched.
- In the world in 1950, I[st] cabbage hybrid was developed in Japan using self incompatible lines and this hybrid was known as Nagaoka No-1
- Gynoecious female sex form in cucumber is controlled by single dominant gene (Acr or F)
- In muskmelon, monoecious and gynoecious lines are produces oblong shape fruits and perfect flower produces round fruits.
- Thinning is compulsory in beet root and turnip
- Dr. Kihara developed Pusa Bedana (seedless variety of watermelon) in 1972 Triploid
- Cercospora leaf spot of sweet potato – 1st reported in Africa
- Temperate vegetables grown in warm season 1. New Zealand spinach 2. Sweet potato
- Spineless variety of Okra - Pusa Sawani
- Guar Gum is extracted from: Cluster bean
- Vine turing is cultural practices followed-Sweet potato
- Type of fruit in Amaranthus-Utricle
- 'Caro-red'beta carotene rich cultivar- Tomato
- Recently discovered new plant species Wild Okra Named "*Abelmoschus Odishae*"
- *Abelmoschus Odishae* is used for hybridization to create better variety with high disease resistant.
- All vegetables come under category spermatophyta
- All vegetables lack fat content (0.1%)
- "Icebox" types cultivar are related-Watermelon

- Vegetable crop recommended for its own iodine rich-Okra
- Ethylene absorabnt- KMnO4
- Mature seeds of Pea and Bean contain Phytic acid
- Inter specific hybrid of Tomato Hissar Anmol Hissar Arun x *L. hirsutum* 2 Rusa Red Plum L. *esculentum x L. Pimpillifolium*
- Ash gourd longest storage life among cucurbits
- Yellow coloured Onion varieties are mostly grown in the world.
- P increases number of flowers in tomato
- Cu increase numbers of flowers in Brinjal
- Zn increases weight of fruit in Brinial
- Little leaf of Brinjal is due to mycoplasma and Zn deficiency
- Carrot asiatic type core is distinct
- European type: core is indistinct, Stump and blunt.
- Varieties of savoy cabbage Perfection Chieftain Wake field, Charlatan Jercey.
- 6 Stamens are present in cruciferae and alliaceae family of vegetables & Petals
- 5 Stamen are present in Solanaceae family of vegetables.
- Among crucifers - Knol khol has highest seed rate-1.5 kg/ha
- Flower colour: Cruciferae Yellow
- Except Chinse kal-White
- Pointed Gourd, Snake Gourd, Bottle Gourd – White flower
- **Total root parasite:** Orobanchae (Broom rape) *i.e Brijal, Chilli, Potato, Mustarad*
- Partial root parasite *i.e* Striga (Witches weed**)**
- The concept of pure line theory was coined by Johanson, 1903 working with Princes var. of Rajma
- Term “Muatation” was coined by Hugo de Varies (1900)
- Non Ionizing radiation of mutagen: UV- rays
- Potato tuber spindle disease caused by viroid
- Fruit weight of spine gourd: 10-20gm
- Following potato cultivar suited for French fry production: Kufri Fryom
- Types of seed germination in pea: Hypogeal
- ‘Kashi Khushi’ an improved cultivar developed by IIVR: Satputia
- ‘Ben oil’ is extracted from: Moringa

- Truss hooking is an polyhouse operation is practices in: Tomato
- CIPHET is situated :Ludhiana
- Turmeric has antioxidant property
- NARS refers to :National Agriculture Research System
- Potato is rich source of :Starch
- Lady bird beetle attack on :Cucurbitaceae family

Growth Regulators

Sr. no.	Common Name	Commercial uses
1.	Auxin	Apical bud dominance, root elongation and promote cell division in tissue culture
	IAA	Prevent premature fruit drop
	IBA	Root formation
	NAA	Fruit thinner, prevent fruit drop
	2,4-D	Fruit setting hormone(<20ppm) and also used as herbicide (>20ppm)
2.	Cytokinin	Dormancy breaker, stimulate cell division
3.	GA_3	Cell elongation, increase fruit size
4.	Ethylene	Ripening of fruit **Ethrel-** Liquid form of ethylene **Ethephone-** Gaseous form of ethylene
5.	CCC Chloromequat (Cycocel)	Lodging preventor
6.	Mallic hydrazide	Growth retardant, prevent onion srouting
7.	6-Benzyl adenine	Senescence delayer

Growth regulators used for promoting female flower production in different cucubits-

Watermelon	TIBA @ 50 ppm, MH @ 50 ppm
Muskmelon	MH @ 200 ppm
Pumpkin, Sponge gourd	Ethrel@250 ppm
Cucumber	Ethrel @ 250 ppm, MH @ 100 ppm
Snake gourd	Ethrel @ 100 ppm
Ridge gourd	Ethrel @ 300 ppm
Summer squash	Ethrel @ 600 ppm
Bottle gourd	MH @ 50 ppm, B @ 3 ppm, Ca @ 5 ppm, PCPA @ 100 - 200 ppm
Bitter gourd	NAA @ 100 ppm, BA @ 25 ppm, Ethrel @ 200 ppm

Maturity Indices in Some Vegetable Crops

- ☆ Tapping- Watermelon
- ☆ Full slip stage- Muskmelon Except Hara Madhu variety (netting)
- ☆ Solidity- Cabbage
- ☆ Juciness- Sweet corn

Different b/w Fruits and Vegetable

Vegetables	Fruits
Most of vegetable are annual/seasonal except chow chow, asparagus	Fruits are perennial except strawberry
All plant parts are use as vegetable	Only fruit are edible part
Generally consuming after cooking	Generally consumed as raw
Vegetable are generally alkaline in nature except tomato, rhubarb	Fruits are generally acidic in nature except strawberry
Vegetables are generally propagated by seed except Pointed gourd, ivy gourd	Mostly fruits are asexually propagated except papaya, acid lime
No training and pruning is practice in vegetable except green house tomato single stem training	Training and pruning is common practice in fruits

Based on methods of raising

Direct sown crops	Transplanted crops
Okra, Carrot, Radish, Beans, Peas, Garlic.	Tomato, Brinjal, Chilli, Cabbage, Cauliflower

Based on Forcing

Cool forcing vegetables	Warm forcing vegetables
Asparagus, Beet, Carrot, Cauliflower, Radish, Spinach, Pea, Onion, Lettuce, Celery	Brinjal, Bean, Tomato, Cucumber, Muskmelon

Based on Soil Factor

Basophiles	Vegetables which grow well in alkali soils
Acidophiles	Vegetables which grow well in acid soils
Neutrophiles	Vegetables which grow well in neutral soils
Psammophytes	Vegetables which grow in soil having sand and gravel
Petrophytes	Vegetables able to grow on rocks

Based on Pigmentation

Green colour vegetables	Red/blue colour vegetables	Yellow colour vegetables
Green colour of leafy vegetables is due to chlorophyll	Red/blue colour of vegetables is due to anthocyanin pigment It is **highly soluble in water**	Yellow colour of vegetables is due to the presence of carotenoids
Under acid condition, chlorophyll changes to olive green in colour	Anthocyanin changes to purple colour under neutral medium, red colour under acidic medium	
In alkali condition, it changes to bright green in colour	blue colour under alkaline medium	

Insecticides

Mode of Action of Insecticides			
Systemic poison (it control sucking pests)	**Contact poison (death after contact)**	**Stomach poison (it control chewing type insects)**	**Fumigant (all types of insects irrespective of its feeding habits)**
Monocrotophos 36SL	Monocrotophos 36SL	Heptachlor	Aluminium phosphide.
Dimethoate	Malathion 50EC	Endosulfan	
Metasystox and Systox	Diazinon	Chlordane	
Dicrotophos	Thiodan	Aldrin	
Phosphomidon	carbaryl	Quinolphos	
Aldicarb	Baygon		
Carbofuran			
Thionazin			
Phorate			

-Safe insecticide of vegetable crops is Malathian

- Dicofol is a miticide
- Chlorpyrifos 20EC is used to control of termite
- Termite control can also done by irrigation

Fungicide

Chemical seed treatment			
Inorganic		**Organic**	
Mercurials	Non mercurials	Mercurials	Non mercurials
Mercuric chloride	Basic copper carbonate	Agrosan GN, Ceresan	Thiram
Mercurous chloride	Red cuprous oxide	Ethyl mercuric chloride	Oxanthine
Carbonate		Phenyl mercuric acetate	Carboxins

Physical methods (Jensen) for the first time used heat to control internal infection of potato tubers by late blight of potato (*Phytophthora infestans*).

- It was observed that heat treatment of potato at 40°C for 4 hours results into death of the internal mycelium.
- Hot water treatment is also effective in the control of nematodes.
- Solar energy treatment to control loose smut was first developed by Luthra.
- Hot air treatment for the control of virus in propagating stocks was first developed by Kunkal in Peach yellow.
- **Mencozeb:** Best widely used foliar spray fungicide.
- Vapan/Methan sodium (soil fumigant) Control of soil fungi and act as herbicide
- **Zineb:** It is used for soil treatment to control damping of and other soil born disease and also meliodogyne
- **Meneb:** control of blight disease of tomato and potato

Weedicide

On the basis of translocation	
Systemic herbicide	Contact herbicide
Propanil, 2,4-D, Atrazine, Simazine.	Paraquat & Diquat

On the basis of time of application	
Pre-plant applied	Fluchloralin and Alachlor
Pre-emergence	Only selective herbicides
Post-emergence	2,4-D, Propanil
On the basis of Selectivity	
Selective herbicide (Such herbicide kills only target weeds)	Non-selective (kills all vegetation (crop + weed both) when comes in contact)
Simazine	Glyphosate
Atrazine	Diquat
2,4-D, MCPA	Paraquat
Butachlor,	
Pendimethalin, ,	
Fluchloralin	
Isoproturon	

- Harvesting stage of vegetable for canning: Tender ripe and firm
- The pungent odorous taste of onion bulb is on account of essential element is: Sulphur
- Vegetables are packed in cams and processed by heat in: Retourt

Tomato product with TSS

Product	TSS
Tomato juice	5%
Tomato soup	7%
Tomato puree	9%
Tomato paste	25%
Tomato kechup	28%
Tomato sauces	30%

Biodiversity Act, 2002

- **National Biodiversity Authority:** Chennai
- **Biodiversity International:** Rome (Italy)
- Geographical Indication of goods (Registration and Protection) Act, 1999
- **Geographical Indication (GI) office:** Chennai
- GI is a sign / mark is attribute to a good of any category (agricultural goods or other industrial/ manufacturing goods etc.)
- Arka Pragati variety of onion is suitable for both *Kharif* and *Rabi* season
- Shape of 'Scarlet Globe' variety of radish is round
- Frequency of spontaneous mutation is 10^{-6}
- **'Sui generis' is latin phrase means:** A system of its own
- Bt. Brinjal developed for resistant to fruit and shoot borer
- Rutabaga is a man made vegetable
- **Seedless watermelon is developed through:** polyploidy
- **Green leafy vegetables are rich sources of:** Folic acid
- **Floral part of saffron is used as spice:** Stigma

CHAPTER-22

Intellectual Property Rights

Choose the correct alternate from given alternatives:

1. **In which year GATT was established**

   (a) 1995 (b) 1970

   (c) 1966 (d) 2001

2. **Which agreement of GATT are related to agriculture**

   (a) Agreement on agriculture (b) TRIPS

   (c) TBT **(d) All the above**

3. **In which year PPV and FR was passed**

   (a) 2001 (b) 2003

   (c) 1999 (d) 2003

4. **Geographical indications of goods (registration and protection) act was passed in year**

   (a) 1999 (b) 2003

   (c) 1970 (d) None of these

5. **National Biological Diversity act was passed in the year**

   (a) 1999 **(b) 2003**

   (c) 1970 (d) 1995

6. **Amendments in patent act was passed in the year**

   (a) 1970 (b) 1966

   (c) 2003 (d) None of these

7. **In which year seed act was passed**

   (a) 1970 (b) 2003

   (c) 1966 (d) None of these

8. **TRIPS under WTO contains how many articles**
   (a) 73 (b) 75
   (c) 85 (d) 7

9. **Under WTO, TRIPS articles are divided into how many parts**
   (a) 7 (b) 5
   (c) 8 (d) 10

10. **WIPO was entered into force in**
   (a) 1970 (b) 1967
   (c) 1995 (d) None of these

11. **Where is the head quarter of UPOV**
   (a) Zenewa (b) Paris
   (c) Bern (d) Zurich

12. **Where is the head quarter of WTO**
   (a) Zenewa (b) Paris
   (c) Bern (d) Zurich

13. **At present how many members of WTO**
   (a) 164 (b) 170
   (c) 180 (d) 185

14. **In which year the patent act was passed**
   (a) 1970 (b) 2000
   (c) 1999 (d) 1957

15. **In which year the new design act was passed**
   (a) 1970 (b) 2003
   (c) 2000 (d) None of these

16. **In which year the new trade mark act was passed**
   (a) 1999 (b) 1970
   (c) 2000 (d) None of these

17. **In which year the copy right act was passed**
   (a) 2000 (b) 2003
   (c) 1957 (d) None of these

18. **In which year copy right rules was passed**
   (a) 1958 (b) 1957
   (c) 2000 (d) None of these

19. Which one is an author

(a) Book writer
(b) Music composer
(c) Film producer
(d) All of these

20. For a patent the invention should satisfy the following criteria

(a) Novelty
(b) Non-obviousness
(c) Usefulness
(d) All of these

21. The patent cooperation treaty (PCT) a multilateral treaty entered in force in the year

(a) 1978
(b) 1980
(c) 1982
(d) None of these

22. In which city the branch offices of the patent office are located in India

(a) New Delhi
(b) Mumbai
(c) Chennai
(d) All the above

23. The registration of GI shall be of how many years

(a) 10
(b) 8
(c) 5
(d) None of these

24. A trade mark may be

(a) Word
(b) Name
(c) Symbol
(d) All of these

25. Who have right in musical sound recording

(a) Lyricist
(b) Composer
(c) Singer
(d) Musician
(e) Producer
(f) All the above

26. What is the maximum duration of plant breeder's right

(a) 18 years
(b) 15 years
(c) Both a and b
(d) None of these

27. What is the maximum duration of farmer's right

(a) 18 years
(b) 15 years
(c) Both a and b
(d) none of these

28. Which of the following are essential requirements for registration of a variety by a plant breeder?

(a) Novelty
(b) Distinctiveness
(c) uniformity
(d) stability
(e) All of these

29. Which of the following are essential requirements for registration of a variety by a farmer?

(a) Novelty (b) Distinctiveness

(c) uniformity (d) stability

(e) All of these

30. Which one is true about the charging of penalty for infringement in case of plant breeder's right?

(a) Charged (b) Not charged

(c) Both a and b (d) None of these

31. Which one is true about the charging of penalty for infringement in case of farmer's right?

(a) Charged **(b) Not charged**

(c) Both a and b (d) None of these

32. If published within the life time of the author of a literary work the term is for the life time of the author plus

(a) 60 years (b) 20 years

(c) 30 years (d) 40 years

33. For cinematography films, records, photographs, posthumous publications, anonymous' publication, works of government and international agencies the term the term of copy right is for how many years from the beginning of the calendar year following the year in which the work was published.

(a) 60 years (b) 20 years

(c) 30 years (d) 40 years

34. For broadcasting the copy right term is for how many years from the beginning of the calendar year following the year, in which the broadcast was made.

(a) 25 years. (b) 20 years

(c) 30 years (d) 40 years

35. In the case of anonymous or pseudonymous works, the copyright is for how many years from the end of the year in which the work is made available to the public with the consent of the owner of the copyright or from the end of the year in which it is reasonable to presume that the author died, which ever term is shorter.

(a) 60 years. (b) 20 years

(c) 30 years (d) 40 years

36. What does copy right cover?

(a) Literary work (b) Artistic work

(c) Cinematographic films (d) Recording

(e) All of these

37. Geographical indications serve the same functions as trademarks, because like trademarks they are:

(a) Source-identifiers (b) Guarantees of quality

(c) Valuable business interests **(d) All of these**

38. The main features of moral write are

(a) Registration is not compulsory

(b) Duration of protection is end with the life of the author

(c) Matter covered as under copy right

(d) All the above

39. What are the right covered under moral right?

(a) Right to paternity (b) Right to integrity

(c) Right to attribution (d) Right to disclosure or privacy

(e) All of these

40. What are the advantages of moral right?

(a) Registration is not compulsory

(b) It promotes creative works

(c) Protection is life long

(d) It provides benefit to the owners

(e) All of these

41. What are the disadvantages of moral right?

(a) It is not transferable

(b) It may lead to monopoly

(c) It does not provide legal right without registration

(d) All of these

42. What is the duration of mask right under registered condition?

(a) 10 years (b) 15 years

(c) 60 years (d) None of these

43. What is the duration of mask right under unregistered condition?

(a) 10 years **(b) 2 years**

(c) 60 years (d) None of these

44. What is the duration of moral right?

(a) 10 years (b) 15 years

(c) 60 years **(d) Life of the author**

45. What is the period of protection of supplementary protection certificate?

(a) 5 years (b) 10 years

(c) Till the life of author (d) None of these

46. Which of the following is the advantage of supplementary protection certificate?

(a) Registration is not compulsory

(b) It promotes research work on pharmaceuticals and pesticides

(c) It extend life span of a patent by 5 years

(d) It provides benefit to the owners

(e) All of these

47. Which of the following is the disadvantage of supplementary protection certificate?

(a) It has short life span

(b) It may lead to monopoly

(c) It does not provide legal right without registration

(d) All of these

48. Which one is correct about transfer of supplementary protection certificate?

(a) Permitted (b) Not permitted

(c) Both a and b (d) None of these

49. Which one is correct about transfer of moral right?

(a) Permitted **(b) Not permitted**

(c) Both a and b (d) None of these

50. Which one is correct about transfer of mask work?

(a) Permitted (b) Not permitted

(c) Both a and b (d) None of these

51. Which one is correct about infringement of moral right?

(a) Possible (b) Not possible

(c) Both a and b (d) None of these

52. Which one is correct about infringement of mask work?

(a) Possible (b) Not possible

(c) Both a and b (d) None of these

53. Which one is correct about infringement of supplementary protection certificate?

(a) Possible (b) Not possible

(c) Both a and b (d) None of these

54. When the supplementary protection certificate come into enforcement ?

(a) After registration (b) After release of copy righted work

(c) Both a and b (d) None of these

55. When the moral right come into enforcement ?

(a) After registration **(b) After release of copy righted work**

(c) Both a and b (d) None of these

56. When the mask work come into enforcement ?

(a) After creation (b) After release of copy righted work

(c) Both a and b (d) None of these

57. Which one is correct about registration of supplementary protection certificate?

(a) Compulsory (b) Not compulsory

(c) Both a and b (d) None of these

58. Which one is correct about registration of moral right?

(a) Compulsory **(b) Not compulsory**

(c) Both a and b (d) None of these

59. Which one is correct about registration of mask work?

(a) Compulsory **(b) Not compulsory**

(c) Both a and b (d) None of these

60. What type of matter covered under supplementary protection certificate?

(a) Medicinal products and plant protection products

(b) Copy righted matters

61. What type of matter covered under moral right?

(a) Medicinal products and plant protection products

(b) Copy righted matters

62. What is correct about the registration of Plant breeder's right?

(a) Essential (b) Not essential

(c) All of these

63. **What is correct about the registration of Plant breeder's right?**
    (a) Essential (b) Not essential
    (c) All of these

Fill in the blanks with appropriate answers

1. Full form of WTO is....**World Trade Organization.**
2. Full form of GATT is....**General Agreement on Tariff and Trade.**
3. Full form of AoA is....**Agreement on Agriculture.**
4. Full form of TBT is....**Technical Barriers to Trade.**
5. Full form of IPR is....**Intellectual Property Right.**
6. Full form of CBD is....**Conventions on Biological Diversity.**
7. Full form of TRIPS is....**Trade Related Aspects of Intellectual Property Right.**
8. Full form of GI is.....**Geographical Indications.**
9. Full form of WIPO is....**World Intellectual Property Organization.**
10. Full form of UPOV is ...**International Union for the Protection of New Varieties of Plants.**
11. Formula of medicine is example of ....**trade secret.**
12. Trade secret can be protected for...... **indefinite period.**
13. Trade secret..... is lost if secret is revealed.
14. Bio piracy ...is the illegal collection of plants by cooperation who patent them for their own use.
15. Domain Name can be registered up to ...**63**... characters long.
16. Full form of DNS is....**Domain Name System.**
17. Full form of LTD is.....**Top Level Domain.**
18. Full form of URL is.....**Uniform Resource Locator.**
19. 'Aspirin' is the example of..... **trade Name.**
20. The duration of protection of plant breeders right for tree is ..**25**...years.
21. The duration of protection of plant breeders right for field crops is ..**25**...years.
22. The distinctive slogan may the ....**trade mark.**
23. A distinctive image may be a ...**trade mark.**
24. A distinctive symbol may be the .... **trade mark.**
25. Registration for plant breeders right is ....essential
26. Registration for farmer's right is ....essential.

27. Duration of registration of explants varieties etc. in case of plant breeder's right is for...**18** years
28. Duration of registration of explants varieties etc. in case of farmer's right is for...**15.** years
29. Plant breeder's right comes into enforcement after ...registration.
30. Farmer's right comes into enforcement after ...registration.
31. Novelty, distinctiveness, uniformity and stability are the essential requirement for registration of a variety in case of .....**plant breeder's**..... right.
32. Novelty, distinctiveness, uniformity and stability are the essential requirement for registration of a variety in case of .....**farmer's**..... right.
33. Registration of a variety in case of farmer's right is valid only in the ....**country** ...of registration
34. Registration of a variety in case of plant breeder's right is valid only in the .... **country** ...of registration
35. For registration of a variety fee is not charged in case of... **farmer's...** right
36. For registration of a variety fee is charged in case of ...**plant breeder's** .. right
37. For annual renewal of registration of a variety fee is not charged in case of ...**farmer's**.. right
38. For annual renewal of registration of a variety fee is charged in case of... **plant breeder's** ..right.
39. Penalty for infringement is charged in case of ...**Plant breeder's**... right.
40. Penalty for infringement is not charged in case of ....**farmer's** ....right.
41. Community right is not covered in case of..**plant breeders**.... right
42. Community right is covered in case of ...**farmer's**... right
43. Authorization is easy in case of... **plant breeder's**... right
44. Authorization is not permitted in case of ...**farmer's** ...right
45. If published within the life time of the author of a literary work the term the term of copy right is for the life time of the author plus ...**60** ....years.
46. For cinematography films, records, photographs, posthumous publications, anonymous' publication, works of government and international agencies the term the term of copy right is .....**60** .....years from the beginning of the calendar year following the year in which the work was published.
47. For broadcasting the copy right term is..... **25**..... years from the beginning of the calendar year following the year, in which the broadcast was made.

48. In the case of anonymous or pseudonymous works, the copyright is for ...**60**..... years from the end of the year in which the work is made available to the public with the consent of the owner of the copyright or from the end of the year in which it is reasonable to presume that the author died, which ever term is shorter.

CHAPTER-23

Model Test Paper

MODEL TEST PAPER-1

1. **The highest cross pollinated among the Cole crops?**
   (a) Cabbage (b) Cauliflower
   (c) Broccoli (d) Kale

2. **Which vegetable crop is day neutral?**
   (a) Tomato (b) Potato
   (c) Sweet potato (d) Spinach

3. **Acta Horticulturae is a publication of?**
   (a) ASHS (b) ISHS
   (c) JSHS (d) HIS

4. **Seed act was passed in India in the year?**
   (a) 1966 (b) 1963
   (c) 1969 (d) 1968

5. **Which of the following factor is mostly maintained under green house?**
   (a) Temperature and humidity (b) Temperature and CO2
   (c) Humidity and ventilation (d) Light and CO2

6. **Silver coloured shade nets are used?**
   (a) To destroy the weeds (b) To control aphids
   (c) To control fungal disease (d) All of theses

7. **Lack of moisture causes**
   (a) Decrease in acidity (b) Decrease in peel thickness
   (c) Decrease in firmness (d) All of these

8. Which vegetable crop is commercially used for processing in India?

(a) Tomato (b) Ash gourd
(c) Tapioca (d) All of these

9. Tetradynamous is a special features among which groups?

(a) Cucurbitaceous (b) Solanaceous
(c) Legume (d) Crucifer

10. The cross of Raphano-Brassica?

(a) Cabbage and radish (b) Turnip and cabbage
(c) Radish and cabbage (d) Cabbage and cauliflower

11. Tomato variety which can with stand high as well as low temperature is

(a) Pusa Naveen (b) Pusa Sadabahar
(c) Pusa Sheetal (d) Hissar Lalit

12. In Tomato red colour is due to

(a) Lycopene (b) Anthocynin
(c) Xanthophylls (d) None of theses

13. Brassica U- Triangle was first given by .

(a) Nieuhwhof (1969) (b) Nagaharu (1935)
(c) Johanson (1935) (d) Willians (1982)

14. Polythene sheet is used as a covering material to prevent

(a) UV rays (b) Humidity
(c) Temperature (d) CO2

15. Yellow anther type variety of chilli is

(a) Bhaskar (b) Jawala Mukhi
(c) Andhra Jyoti (d) Bhagyalakshmi

16. The red colour of chilli fruit at ripening stage is due to pigment

(a) Capsacin (b) Capsicutin
(c) Capsanthin (d) Lycopene

17. Massing and Selfing technique of breeding was first time suggested in

(a) Potato (b) Onion
(c) Turnip (d) Non

18. Oleoresin in extracted from

(a) Colocasia (b) Chilli
(c) Brinjal (d) Onion

19. Mostly used protected structure by nurserymen in India is

(a) Plastic tunnel (b) Poly house
(c) Net house (d) Glass house

20. Ex-situ conservation restricts

(a) Gene flow (b) Inbreeding
(c) Disease and pest (d) All of these

21. Recently commercial bio-agent/bio –fertilizer is

(a) Phosphorus solubilizing (b) Potassium solubilizing
(c) VAM (d) Rhizobium

22. Meristem tip culture is done to avoid

(a) Insect infestation (b) Virus elimination
(c) Fungal infection (d) Bacterial infection

23. No.1 processing vegetable in the world is

(a) Brinjal (b) Onion
(c) Chilli (d) Tomato

24. Seed rate of chilli per hectare is

(a) 0.5 kg/ha (b) 1.5 kg/ha
(c) 2.5 kg/ha (d) 3.5 kg/ha

25. Which of the following is leading vegetable crop in India in area as well as production

(a) Cabbage (b) Potato
(c) Tomato (d) Brinjal

26. Round type variety of Brinjal is

(a) Arka Shirish (b) Arka Kusumkar
(c) Pusa Purple Cluster (d) Punjab Bahar

27. In our diet, staple vegetable is

(a) Brinjal (b) Potato
(c) Tomato (d) Chilli

28. The Asveg-1 is a tropical hybrid among which crop?

(a) Red cabbage (b) Brussels sprout
(c) Chinese cabbage (d) Kale

29. Which of the following statement is correct regarding onion seed production?

(a) Day length is more important than temperature

(b) Temperature is more important than day length

(c) Temperature and day length have equal importance

(d) None of the statements are correct

30. Flower colour of onion is

(a) Purple
(c) Red
(b) Yellow
(d) White/blue

31. Massing and Selfing technique of breeding was first time suggested in

(a) Papaya
(b) Turnip
(c) Allium spp.
(d) Mango

32. Resistance source, for YVMV is

(a) A. tuberculatus
(b) A. Manihot
(c) A. tetraphyllus
(d) A. esculantus

33. Chromosome number of cultivated Okra is

(a) 2n = 58
(b) 2n = 194
(c) 2n = 130
(d) None of the above

34. Parentage of Pant C1 variety of chilli is

(a) NP 46 A × Puri Red
(b) NP 46 A × Kandhari
(c) G4 × Anther mutant
(d) Pusa Jwala × IC 31339

35. The crop showing greatest inbreeding depression is

(a) Onion
(b) Lettuce
(c) Tomato
(d) Lima bean

36. Drought tolerant onion cultivar is

(a) Hissar-2
(b) Pusa Ratnar
(c) Pusa Red
(d) Baswant -780

37. Transverse cotyledon cracking is a major problem in

(a) Pea
(b) French bean
(c) Cowpea
(d) Pumpkin

38. Bud necrosis of watermelon is a viral disease, which is transmitted by

(a) Aphids
(b) Thrips
(c) White fly
(d) Jassids

39. One of the progenitor of cultivated okra is

(a) Abelmoschus crinitus (b) Abelmoschus angulosus

(c) Abelmoschus caillei (d) Abelmoschus tuberculatus

40. 'White heart' is a physiological disorder of

(a) Carrot (b) Beet root

(c) Muskmelon (c) Watermelon

41. The desirable acidity in tomato should be

(a) 0.1% (b) 0.2%

(c) 0.3% (d) 0.4%

42. Potato leaf marker commonly associated with male sterility line of

(a) Tomato (b) Brinjal

(c) Chilli (d) Capsicum

43. Pre harvest spray of which of the following is done to reduce storage loss of onion ?

(a) Maleic hydrazide (b) Gibberellic Acid

(c) Indole butyric Acid (d) Ethephon

44. Selfing variety of onion is

(a) Pusa White Flat (b) Pusa Madhvi

(c) Pusa Red (d) Arka Pragati

45. Chinese cabbage can first to be harvested after sowing.

(a) 30 days (b) 40 days

(c) 50 days (d) 60 days

46. The chromosome number in Chinese cabbage?

(a) 2n=18 (b) 2n=22

(c) 2n=20 (d) 2n=16

47. Parentage of Pusa Jawala variety of chilli is

(a) NP 46 A × Puri Red (b) NP 46 A × Kandhari

(c) G 4 × Anther mutant (d) G4 × Yellow Anther mutant

48. 'Blossom end rot' of tomato is due to deficiency of

(a) Mg (b) Bo

(c) Ca (d) Mo

49. Which fungicide is most effective for controlling powdery mildew of pea crop

(a) Captan (b) Mancozeb

(c) Karathane (d) Thiram

50. Homeostasis operates in the resistance mechanism known as

(a) Active resistance (b) Field resistance

(c) Horizontal resistance (d) Vertical resistance

Answer

1. (c)	2. (a)	3. (b)	4. (a)	5. (b)	6. (b)	7.(d)	8.(a)	9.(d)	10.(c)
11. (b)	12. (a)	13. (b)	14. (a)	15. (a)	16. (c)	17.(b)	18.(b)	19.(a)	20.(d)
21. (b)	22. (b)	23. (d)	24. (b)	25. (b)	26. (d)	27.(b)	28.(c)	29.(b)	30.(d)
31. (c)	32. (b)	33. (c)	34. (b)	35. (a)	36. (a)	37.(b)	38.(b)	39.(d)	40.(c)
41. (c)	42. (d)	43. (a)	44. (b)	45. (b)	46. (c)	47.(a)	48.(c)	49.(c)	50.(c)

MODEL TEST PAPER-2

1. **Cat face is a disorder of**
   (a) Cauliflower (b) Pointed gourd
   (c) Tomato (d) chow-Chow

2. **Which system is used for hybrid seed production in onion**
   (a) Self-incomoatibility (b) Genetic male sterility
   (c) Cytoplasmic male sterility (d) All of these

3. **Which of the following is a miticide (acaricides)?**
   (a) DDT (b) Carbofuran
   (c) Chloropyrifos (d) Dicophol

4. **Herbicide should not be applied in onion at**
   (a) Emergence-post crook stage (b) Emergence-loop stage
   (c) Emergence-crook stage (d) Loop-post crook

5. **Acidic soil encourages**
   (a) Clubroot in crucifers (b) Black leg in crucifers
   (c) Verticillium wilt in solanaceous (d) Root knot nematode

6. **Which of the following is associated with vegetable research?**
   (a) Swaminathan (b) G. Kallo
   (c) T.K. Bose (d) Randhawa

7. **The stalk of the ovule is called**
   (a) Panicle (b) Micopyle/ Funicle
   (c) Pedicel (d) Petiole

8. **Genetically modified crop recommended for commercial cultivation in india is**
   (a) Tomato (b) Potato
   (c) Brinjal (d) Chilli

9. **Arka Navneet is the variety of**
   (a) Tomato (b) Potato
   (c) Brinjal (d) Chilli

10. **Origin of Potato is**
    (a) North America (b) South America
    (c) China (d) India

11. Wart disease of Potato is caused due to

(a) Virus (b) Bacteria

(c) Fungus (d) Mycoplasma

12. Among which is regarded as mini cabbage/miniature cabbage .

(a) Broccoli (b) Brussels sprout

(c) Kale (d) Chinese cabbage

13. Jade cross (F1) is a famous hybrid which crop?

(a) Cabbage (b) Cauliflower

(c) Brussels sprout (d) Kale

14. Hollow heart in Potato is due to

(a) Moisture deficiency (b) Poor ventilation

(c) Excessive N2 (d) Low temperature

15. Black leg of cabbage is caused by

(a) Fungus (b) Bacteria

(c) Virus (d) MLO's

16. In which state Potato is generally not grown

(a) Andhra Pradesh (b) Kerala

(c) Uttar Pradesh (d) Madhya pradesh

17. Pungency of Garlic is due to

(a) Allyl propyl disulphide (b) Diallyl disulphide

(c) Isothiduinate (d) None of these

18. Based on the respiration rate cabbage are comes under category

(a) Very high (b) High

(c) Medium (d) Low

19. Which molecular marker reported to be linked DGMS in cabbage?.

(a) RAPD (b) RFLP

(c) AFLP (d) SSR

20. Root of radish is

(a) Conical form (b) Napiform

(c) Fusiform (d) None of these

21. Major glucosinolates present in radish is

(a) Isothiocyanates (b) Allyl isothiocyanates

(c) Both a & b (d) None of theses

22. Removal of central portion of curd for easy initiation of flower stalk in cauliflower is called as

(a) Topping (b) Scooping
(c) Blanching (d) All of the above

23. Arka Abir is the variety of

(a) Brinjal (b) Chilli
(c) Tomato (d) Cucurbit

24. Common practice in cauliflower for protect curd from yellow colour is

(a) Scooping (b) Earthing up
(c) Blanching (d) None of these

25. Chromosome number of turnip is

(a) 2n = 18 (b) 2n = 20
(c) 2n = 22 (d) 2n = 24

26. The seed rate of Brinjal (gm/ha) is

(a) 400-500 (b) 300-400
(c) 100-200 (d) 200-300

27. Polyphenol oxydase enzyme inactivate at the temperature of

(a) 80-90°C (b) 100°C
(c) 50°C (d) above 100°C

28. Turnip is botanically modified form of root is called as

(a) Napiform (b) Fusiform
(c) Conical form (d) All of these

29. Beetroot belongs to which family

(a) Brassicaceae (b) Chenopodiaceae
(c) Apiaceae (d) Basillaceae

30. Folic acid is essential for pregnant women to reduce risk of spina bifida, which is predominantly found in

(a) Beetroot (b) Radish
(c) Turnip (d) Carrot

31. For biological control to tomato fruit borer, the ideal trap crop is:

(a) Tobacco (b) Mustard
(c) Garlic (d) Marigold

32. Seed rate of tomato (gm/ha) is: 11

(a) 400-500 (b) 800-1000
(c) 1200-1500 (d) 100-200

33. The 1000 seed weight of cabbage is

(a) 3. 3 g (b) 2.8 g
(c) 5.5 g (d) 4. 5 g

34. The SMR (Seed multiplication rate) of cabbage and cauliflower?

(a) 50 (b) 100
(c) 200 (d) 350

35. The pungency of chilli is due to

(a) Capsanthin (b) Capsacin
(c) Isothiocynate (d) None of these

36. How many days required from flowering to pod maturity in cauliflower is.

(a) 60-70 day (b) 70-90 day
(c) 60-80 day (d) 90-100 day

37. Andromonecius sex form is present in

(a) Watermelon (b) Muskmelon
(c) Longmelon (d) Round melon

38. Optimum temperature required for vernalization in temperate vegetables

(a) <4°C (b) <7°C
(c) <10°C (d) <15°C

39. For keeping of sugar content low, potato tubers should be stored at a temperature of

(a) 5°C (b) 10°C
(c) 15°C (d) None of these

40. When protoandry and protogyny both are present in the same crop, this is known as

(a) Dudichogamy (b) Heterodichogamy
(c) Homogamy (d) PDSD

41. Which of the following is major problem in green house cultivation

(a) Nematode (b) White fly
(c) Aphid (d) All of these

42. Main aim of blanching is

(a) To remove field heat (b) To inactivate enzyme

(c) To kill microorganism (d) None of these

43. Which of the following is a non PCR based molecular marker

(a) RAPD (b) AFLP

(c) SSR (d) RFLP

44. Anti-cancer property of cabbage is due to presence of

(a) Glucosinolates (b) Sinigrin

(c) Indole-3-carbinol (d) Sulphoraphane

45. Cabbage yellow (Types-A) resistance was governed by......gene.

(a) Single recessive (b) Single dominant

(c) Recessive poly (d) Polygene

46. "Rutin" and Vitamine Pare associated which vegetable crop ?

(a) Chilli (b) Tomato

(c) Brinjal (d) Potato

47. PK-ft-16 is the variety of

(a) Brinjal (b) Chilli

(c) Tomato (d) Cucurbit

48. Which of the following country has highest productivity of vegetable in the world?

(a) Spain (b) Indonasia

(c) USA (d) Thailand

49. Which of the following is a very shallow rooted vegetable

(a) Spinach (b) Beet root

(c) Cabbage (d) Onion

50. Calcium deficiency symptoms are first occurs at

(a) Older leaves (b) Younger leaves

(c) Terminal bud (d) All of these

Answer

1. (c)	2. (c)	3. (d)	4. (a)	5. (a)	6. (b)	7.(b)	8.(c)	9.(c)	10.(b)
11. (c)	12. (b)	13. (c)	14. (b)	15. (a)	16. (b)	17.(b)	18.(d)	19.(a)	20.(c)
21. (a)	22. (b)	23. (b)	24. (c)	25. (b)	26. (b)	27.(a)	28.(a)	29.(b)	30.(a)
31. (d)	32. (a)	33. (a)	34. (b)	35. (b)	36. (b)	37.(b)	38.(b)	39.(b)	40.(b)
41. (a)	42. (b)	43. (d)	44. (c)	45. (b)	46. (a)	47.(b)	48.(a)	49.(b)	50.(c)

MODEL TEST PAPER-3

1. **Progenitor of chilli is**
   (a) C. minimum (b) C. baccatum
   (c) C. fruitesence (d) C. eximinum
2. **Pusa, Sawani variety of bhendi is cross between**
   (a) IC-1542 × Pusa Mukhmali
   (b) Pusa Mukhmali × IC-1542
   (c) A. esculantus × A. manihot spp. Manihot
   (d) A. esculantus x A. tetraphyllus
3. **Which coloured onions are mostly preferred by Indians?**
   (a) White (b) Yellow
   (c) Orange (d) Red
4. **Skin colour of 'Udaipur 102' variety of onion is**
   (a) White (b) Yellow
   (c) Orange (d) Red
5. **Which of the following hormone increase drought tolerance**
   (a) IAA (b) ABA
   (c) Ethylene (d) GA3
6. **Haploid plants can be obtained through**
   (a) Anther culture (b) Meristem culture
   (c) Ovule culture (d) Ovary culture
7. **The .... is done to obtained virus free plants**
   (a) Anther culture (b) Meristem culture
   (c) Cell culture (d) Embryo culture
8. **Critical day length required for bulbing in garlic is**
   (a) 8 hrs. (b) 10 hrs
   (c) 12 hrs. (d) 14 hrs.
9. **Which type of dichogamy is present in carrot**
   (a) Protoandry (b) Protogyny
   (c) Both (d) None of these

10. Which of the following is a cross pollinated bean

(a) Phaseolus vulgaris (b) Phaseolus coccineus

(c) Phaseolus lanatus (d) Phaseolus acutifolius

11. Early Grano variety of onion introduced from

(a) USA (b) Philippines

(c) UK (d) Holland

12. Brown anther male sterility found in

(a) Tomato (b) Onion

(c) Carrot (d) Cauliflower

13. The largest producer of broccoli in world?

(a) India (b) Japan

(c) USA (d) China

14. Yellowing is major storage problem crop.

(a) Cauliflower (b) Cabbage

(c) Brussels sprouts (d) Broccoli

15. The word of broccoli derived from language

(a) Latin (b) German

(c) Italian (d) English

16. Which vegetable crop mostly affected by thrips

(a) Tomato (b) Okra

(c) Chilli (d) Onion

17. Which is a rich sourse of Vit-C

(a) Cabbage (b) Tomato

(c) Chilli (d) All

18. CPRI is situated at:

(a) Lucknow (b) Shimla

(c) Dehradun (d) Pune

19. Potato seed rate (TPS) per ha is

(a) 1g (b) 5g

(c) 10g (d) 15g

20. Pollen sterility in tomato can be identified by using

(a) Acetocarmine (b) KMno4

(c) Methyle blue (d) Phenolphthalein

21. 'Big bud' is a disease of tomato caused by

(a) Fungus (b) Bacteria

(c) Virus (d) MLO'S

22. Seed rate of onion for Rabi season

(a) 12-15 kg/ha (b) 10-12 kg/ha

(c) 25 kg/ha (d) None

23. Punjab Padmini variety of bhendi is cross between

(a) IC-1542 x Pusa Mukhmali

(b) Pusa Mukhmali × IC-1542

(c) A. esculantus x A. manihot spp. Manihot

(d) A. esculantus × A. tetraphyllus

24. Which disorder of tomato associated with improper pollination & fertilization?

(a) Puffiness (b) Cat face

(c) BER (d) Unfruitfulness

25. Which of the following is an acid tolerant in vegetable?

(a) Sweet potato (b) Okra

(c) Brinjal (d) Potato

26. Tomato variety 'Money Maker' is introduced from

(a) Italy (b) France

(c) USA (d) England

27. The fruits of tomato

(a) Stone (b) Drupe

(c) Berry (d) Pome

28. Pusa Ruby is a variety of

(a) Potato (b) Tomato

(c) Brinjal (d) Chilli

29. Leucinodis orbanalis is the major pest of

(a) Tomato (b) Brinjal

(c) Potato (d) Chilli

30. Azad Nutan is a variety of

(a) Tomato (b) Brinjal

(c) Chilli (d) Bottle gourd

31. Pusa hybrid-2 is variety of cauliflower developed was developed by utilized the following breeding method is

(a) Single cross (b) Top cross

(c) Double cross (d) Three way cross

32. Head shape index of cabbage.......ranged.

(a) Normal shaped head = 0. 8-1 (b) Conical head = 1

(c) Drum head = <0. 5 (d) All of these

33. Black rot in cabbage was mainly governed by.....nature.

(a) Mongenic dominant (b) Monogenic recessive

(c) Polygenic resistance (d) Two complementary gene

34. Salt concentration in pickle is measured by

(a) Refractometer (b) Solometer

(c) Salt meter (d) pH meter

35. Richman Long effect describes one of the basic properties of which plant growth regulator

(a) Cytokinin (b) Jasmonic acid

(c) Ethylene (d) ABA

36. Markers used for determining genetic variation in plants is

(a) Biochemical markers (b) Morphological markers

(c) Molecular markers (d) All of these

37. PAR wave length range is

(a) 0.1 to 0.3 μm (b) 0.4 to 0.7 μm

(c) 0.7 to 0.1 μm (d) 1.0 to 1.3μm

38. Pok choi is a .... vegetable

(a) Leafy (b) Flower

(c) Stem (d) Root

39. Cocopeat has ....% porosity

(a) 15-20 % (b) 20-25%

(c) 25-30% (d) 30-35%

40. Strongest self- incompatibility was reported among which Cole crops .

(a) Brussel sprout (b) Cabbage

(c) Indian cauliflower (d) Kale

41. Triple cross hybrid in kale was first proposed by .

(a) Thompson (1964) (b) G.H. Shull (1914)

(c) D.F. Jones (1965) (d) S.J. Stadler (1955)

42. Which among is the leutin rich vegetables .

(a) Brussels sprout (b) Kale

(c) Kohlrabi (d) Chinese cabbage

43. Kitchen kale are belong which following sub. Spp .

(a) Fruticosa (b) Fimbriata

(c) Medullosa (d) Acephala

44. Red colour in onion is due to the pigment

(a) Carotene (b) Quercetin

(c) Lycopene (d) Anthocyanin

45. Red colour in tomato is due to

(a) Anthocyanin (b) Lycopene

(c) Carotene (d) Prolycopene

46. Exposure of potato tuber to sun light causes

(a) Black heart (b) Hollow heart

(c) Browning (d) Greening

47. The economic threshold level (ETL) of aphid population in case of potato is

(a) 20/100 leaves (b) 3/100 leaves

(c) 0/100 leaves (d) 40/100 leaves

48. India's rank in world vegetable production is

(a) First (b) Second

(c) Third (d) Fourth

49. Curcumim is extracted from

(a) Cucumber (b) Capsicum

(c) Ginger (d) Turmeric

50. Multistyle flowers is common in

(a) Okra (b) Brinjal

(c) Chillies (d) Lima Bean

Answer

1. (a)	2. (a)	3. (d)	4. (a)	5. (b)	6. (a)	7.(b)	8.(c)	9.(a)	10.(b)
11. (a)	12. (c)	13. (c)	14. (d)	15. (c)	16. (c)	17.(c)	18.(b)	19.(b)	20.(a)
21. (d)	22. (a)	23. (c)	24. (a)	25. (d)	26. (c)	27.(c)	28.(b)	29.(b)	30.(b)
31. (b)	32. (d)	33. (b)	34. (b)	35. (a)	36. (c)	37.(b)	38.(a)	39.(c)	40.(d)
41. (a)	42. (b)	43. (b)	44. (d)	45. (b)	46. (d)	47.(a)	48.(b)	49.(d)	50.(b)

MODEL TEST PAPER-4

1. **Orgin of Okral is**
   (a) India (b) China
   (c) Africa (d) America

2. **Vegetable included in solanaceous fruit group**
   (a) Tomato (b) Potato
   (c) Sweet potato (d) Okra

3. **The broccoli is mainly processed in form**
   (a) Canned (b) Freezing
   (c) Pickled (d) Drying

4. **Bright green hypocotyl is used as a morphological marker.....crop.**
   (a) Cauliflower (b) Broccoli
   (c) Cabbage (d) Kohlrabi

5. **The ideal knob weight of Kohlrabi...(gm)**
   (a) 100-150 gm (b) 200-250 gm
   (c) 300-350 gm. (d) 400-450gm

6. **'Pusa virat' is an improved variety of which crop .**
   (a) Cabbage (b) Brussels sprout
   (c) Kohlrabi (d) Broccoli

7. **Which variety of tomato is also called as 'cold set'**
   (a) Pusa Sadabahar (b) Pusa Sheetal
   (c) Pusa Uphar (d) None of these

8. **Hybrid variety of brinjal most suitable for Bharta making is**
   (a) BH-2 (b) BH-1
   (c) Pusa Anmol (d) None of these

9. **Little leaf disease of brinjal is caused by**
   (a) MLOs (b) fungi
   (c) Bacteria (d) None of these

10. **Chillies are harvested at for caning purpose**
    (a) Green stage (b) Red stage
    (c) Yellow stage (d) None of these

11. Chilli leaf curl virus is transmitted by

(a) White flies (b) Leaf hoppers

(c) Thrips (d) Mites

12. Cytodeme concept (Harberd, 1972) was commercially utilized in

(a) Cucurbits (b) Legumes

(c) Solanaceous (d) Crucifers

13. Ancestor of garlic is

(a) Allium schoenoprasum (b) Allium longicupis

(c) Allium fistulosum (d) Allium vavilovii

14. Downward curling in chilli is due to attack of

(a) Thrips (b) Mites

(c) White flies (d) Aphids

15. Cole crops are rich source of mineral......

(a) Calcium (b) Phosphorus

(c) Sodium (d) Iron

16. Cole crops are excellent preventive against which types of cancer......

(a) Prostate (b) Bowl

(c) Lung (d) Stomach

17. Pusa Madhvi variety of onion is suitable for cultivation in

(a) Kharif (b) Rabi

(c) Both Kharif and Rabi (d) Zaid

18. Cassava belong to the family

(a) Araceae (b) Euphobiaceae

(c) Asteraceae (d) Poaceae

19. Watermelon bud necrosis is caused by

(a) Fungus (b) Bacteria

(c) Virus (d) Algae

20. Ploidy level in Sweet potato

(a) Diploid (b) Tetrapliod

(c) Hexaploid (d) Octaploid

21. The botanical Name ofPakchoi?

(a) Brassica pekinensis (b) Brassica chinensis

(c) Brassica nigra (d) Brassica carinata

22. Black rot resistance was governed by......gene .

(a) Single recessive (b) Single dominant

(c) Two complementary (d) Poly

23. Pusa Early Bunching is the variety of

(a) Tomato (b) Fenugreek

(c) Carrot (d) Cauliflower

24. Which of the following is highly salt tolerant

(a) Ash gourd (b) Bottle gourd

(c) Bitter gourd (d) Pumpkin

25. Pinching of male flower is effective in hybrid seed production of

(a) Ash gourd (b) Bottle gourd

(c) Bitter gourd (d) Pumpkin

26. Pectin is basically a polymer of

(a) Amino acid (b) D-galactocurnic acid

(c) Gallic acid (d) None of these

27. Sauerkraut is prepared from

(a) Cabbage (b) Cauliflower

(c) Carrot (d) Pumpkin

28. Xanthosoma sagittifolium is a

(a) Root vegetables (b) Tuber vegetables

(c) Bulb vegetables (d) Stem vegetables

29. Which type of cucurbits are planted in polyhouse

(a) Andromonoecious (b) Monoecious

(c) Gynoecium with parthenocarpy (d) Gynoecious

30. The most favourable range of soil pH for tomato cultivation

(a) 4.5-5.5 (b) 6.0-7.0

(c) 7.5-8.5 (d) 7.2-8.0

31. Edible part of Asparagus

(a) Roots (b) Tender shoots

(c) Fruits (d) Flowers

32. Core intact is one of the seed production methods used in vegetable of following crops

(a) Cabbage (b) Cauliflower

(c) Broccloi (d) Brussels sprout

33. Pear shaped fruit variety of tomato is

(a) Punjab Chuhara (b) Punjab Kesari

(c) Punjab Tropic (d) TNR-7

34. Orgin of tomato is from

(a) Mexico (b) Chile

(c) Peru Ecuador (d) India

35. Which method can be employed to maintained self-incompatible inbred lines?

(a) Bud pollination (b) Open pollination

(c) Cross pollination (d) Compatible pollination

36. 'Scooping' is an important operation in raising seed crop in which of the following vegetables crop

(a) Cabbage (b) Cauliflower

(c) Broccloi (d) Brussels sprout

37. Which of the botanical varieties of the lettuce have excellent shipping and handling abilities

(a) Butter head type (b) Crips head type

(c) Cos type (d) Celery lettuce

38. In case of brinjal the phomopsis blight resistant variety is

(a) Pusa Ankur (b) Puas Bhairav

(c) Arka Sheel (d) Black Beauty

39. Which one is used as a donor for brinjal fruit and shoot borer

(a) Solanum nigrum (b) S. melongena

(c) S. Gilo (d) S. muricatum

40. Name of onion variety which in picking type and suitable for export

(a) ADR (b) ALR

(c) Agrifound Rose (d) Agrifound Red

41. In which crop, Gynoecy sex is commercially exploited for hybrid seed production

(a) Watermelon (b) Cucumber

(c) Muskmelon (d) Ridge gourd

42. Pollination and fertilization before anthesis is called

(a) Cleistogamy (b) Homogamy

(c) Chasmogamy (d) Dichogamy

43. The hybrids of crop, mostly preferred in India are

(a) Brinjal (b) Onion

(c) Tomato (d) Cucumber

44. Which crop is a good source of L-Dopa (Dopamine) to cure Parkinson disease in human

(a) Cabbage (b) Cauliflower

(c) Broccloi (d) Brussels sprout

45. Marsh spot in pea occurs due to deficiency of

(a) Mg (b) Mn

(c) B (d) Fe

46. 'Shoe string' is a viral disease of tomato caused by

(a) Strain of TMV (b) Strain of TLCV

(c) Strain of TSWV (d) Strain of CMV

47. Oil content in the okra seed is

(a) 30% (b) 24%

(c) 40% (d) None of the above

48. Brinjal belongs to the species

(a) Melongena (b) Khasianum

(c) Sisymdrifolium (d) None of these

49. Krishna is the variety of

(a) Tomato (b) Potato

(c) Chilli (d) Brinjal

50. "Capsaicin" is chemically

(a) C17H27NO3 (b) C11H22NO3

(c) C18H28NO3 (d) None of these

Answer

1. (c)	2. (a)	3. (b)	4. (b)	5. (b)	6. (c)	7.(b)	8.(a)	9.(a)	10.(b)
11. (a)	12. (d)	13. (b)	14. (b)	15. (b)	16. (b)	17.(c)	18.(b)	19.(c)	20.(c)
21. (b)	22. (b)	23. (b)	24. (a)	25. (b)	26. (b)	27.(a)	28.(b)	29.(c)	30.(b)
31. (b)	32. (a)	33. (a)	34. (c)	35. (a)	36. (b)	37.(b)	38.(b)	39.(c)	40.(c)
41. (b)	42. (a)	43. (c)	44. (c)	45. (b)	46. (d)	47.(c)	48.(a)	49.(d)	50.(a)

MODEL TEST PAPER-5

1. **The optimum size of potato tuber for planting is**
   (a) 10-15g (b) 30-40g
   (c) 750g (d) 100g

2. **Recalcitrant is related with**
   (a) Seed germination (b) Seed dormancy
   (c) Seed storage (d) Seed production

3. **High C:N ratio favours**
   (a) Reproductive growth (b) Vegetative growth
   (c) Bud dormancy (d) All of these

4. **Which of the following breeding method is not suitable for asexually reproducing crops**
   (a) Clonal selection (b) Pedigree breeding
   (c) Mass selection (d) Mutation breeding

5. **The thickness of covering material in poly house should be:**
   (a) 100μ (b) 150μ
   (c) 200μ (d) 300μ

6. **The breeding method used for self pollinated crops**
   (a) Introduction (b) Pure line selection
   (c) Mass selection (d) Clonal selection

7. **Potato is which type of Plant for its tuber**
   (a) Long day (b) Short day
   (c) Day neutral (d) None of these

8. **Mosaic in chilli is transmitted by**
   (a) Hopper (b) White fly
   (c) Aphid (d) None of these

9. **Which of the following crops is produced maximum in the world ?**
   (a) Mango (b) Tomato
   (c) Grape (d) Potato

10. **Late blight resistant variety of potato is**
    (a) K. Chandramukhi (b) K. Jyoti
    (c) K. Alankar (d) K. Badshah

11. The breed for disease resistance which method is used ?

(a) Back cross
(b) Pure line selection
(c) Mass selection
(D) All of these

12. Anther mutant cultivar of tomato is

(a) Pusa Divya
(b) Pusa Gavrav
(c) Pusa Ruby
(d) None

13. "Golden Flake" is due to excess Ca oxalate and low K : Ca ratio in

(a) Tomato
(b) Carrot
(c) Potato
(d) Cabbage

14. Stamenless mutant is reported in :

(a) Cabbage
(b) Brinjal
(c) Tomato
(d) Carrot

15. Self-pollination increase

(a) Homozygosity
(b) Heterozygosity
(c) Homogeneity
(d) Heterogeneity

16. The chromosome number (2n) of French bean is

(a) 18
(b) 20
(c) 22
(d) 24

17. Potato is transformed with protein synthesizer gene from

(a) Fenugreek
(b) Spinach
(c) Beet root
(d) Amaranthus

18. Synthetic seed in vegetables in produced through

(a) Ovule culture
(b) Callus culture
(c) Meristem culture
(d) Somatic embryogenesis

19. Commonly used sprout suppressant in potato is

(a) EDB
(b) Thiourea
(c) CIPC
(d) Biphenyl

20. The cleistogamous vegetable is :

(a) Garden pea
(b) Lettuce
(c) Capsicum
(d) Tomato

21. Which of the following is an ancestor species of watermelon ?

(a) Citrullus vulgaris
(b) Citrullus colocynthi
(c) Citrullus ecirrhosus
(d) Citrullus naudinianus

22. Red colour of onion is due to presence of

(a) Lycopene (b) Anthocynin

(c) Quercetin (d) Xanthophyll

23. Which one var. of tomato sets fruit at low temperature

(a) Pusa Ruby (b) Pant Bahar

(c) Hissar Arun (d) Pusa Sheetal

24. Tomato is a rich source of

(a) Vitamin A (b) Iron

(c) Phosphorus (d) Vitamin C

25. Which of the following species of brinjal is used in grafting

(a) Solanum sisymbriifolium (b) Solanum torvum

(c) Solanum gilo (d) Solanum khasianum

26. Which of the following Solanum species is tolerant to drought

(a) Solanum peruvianum (b) Solanum cheesmani

(c) Solanum galapagense (d) Solanum penelli

27. Which of the following species of tomato is resistant to insect/pest

(a) S. chilense (b) S. hirsutum]

(c) S. cheesmani (d) S. perviflorum

28. Which of the following is/are indigenous vegetable (s)?

(a) Pointed gourd (b) Lablab bean

(c) Brinjal (d) All of these

29. Production of lycopene in tomato hampers adversely

(a) >20°C (b) >25°C

(c) >35°C (d) >30°C

30. 'Gold Fleck' is due to excess of calcium oxalate and low K:Ca ratio occurs in

(a) Potato (b) Carrot

(c) Cabbage (d) Tomato

31. Which of the following crop has 2n=24

(a) Ash gourd (b) Carrot

(c) Cabbage (d) Tomato

32. Ty-1 gene is associate with

(a) Tomato (b) Carrot

(c) Cabbage (d) Potato

33. Diocey is very common in

(a) Asparagus (b) Tulip

(c) Cabbage (d) Potato

34. ELISA test is done to find out ..... diseases

(a) Fungal (b) Bacterial

(c) Viral (d) All of these

35. Seed viability is determined by

(a) Direct germination (b) Excised embryo

(c) Tetrazolium test (d) All of these

36. Which of the following breeding method is most successful in orNamental plants

(a) Pure line selection (b) Mass selection

(c) Mutation (d) Hybridization

37. Which of the following is a strangest senescence stimulator

(a) BA (b) ABA

(c) AOA (d) PBA

38. Normal relative humidity in green house should be

(a) 20-40% (b) 40-60%

(c) 60-80% (d) 80-100%

39. Inheritance of pigments is controlled by .... gene action

(a) Additive (b) Dominance

(c) Epistasis (d) None of these

40. Mutation is a sudden heritable change in the genetic makeup of an organism. It's a

(a) Multicellular event (b) Organ event

(c) Single cell event (d) All of the above

41. Coriander is harvested at which at

(a) Yellowing of leaves (b) Drying of seeds

(c) 50% seeds gets brown (d) None of these

42. The variation seen in plants that's have been produced by plant tissue culture are known as

(a) Chimeric variation (b) Somaclonal variation

(c) Mutated variation (d) None of these

43. In tomato Hybrid Hissar Anmol the donor for leaf curl virus is

(a) L. peruvianum (b) L. esculantum

(c) L. pimpinellifolium (d) L. hirsutum L sp. Glaberatum

44. Which of the following is responsible for cracking in tomato

(a) Ca deficiency (b) B deficiency

(c) S deficiency (d) Mn deficiency

45. Early blight of tomato is caused by

(a) Fungus (b) Virus

(c) Bacterium (d) None of these

46. Little leaf of Brinjal is transmitted by

(a) Hopper (b) Aphids

(c) Jassids (d) White fly

47. Which of the following are phomopsis blight resistant varieties of brionjal ?

(a) Pant Samrat (b) Puas Purple Cluster

(c) Pusa Bhiarav (d) All of these

48. Tomato belongs to the species

(a) peruvianum (b) esculantum

(c) hirsutum (d) pimpenillifolium

49. Which vegetable is the rich source of Vit. C (>100 mg/100g)

(a) Tomato (b) Turnip

(c) Cabbage (d) Radish

50. Cleistogamy favours

(a) Self pollination (b) Cross pollination

(c) Both (d) None

Answer

1. (b)	2. (c)	3. (a)	4. (c)	5. (c)	6. (b)	7.(b)	8.(c)	9.(c)	10.(d)
11. (b)	12. (a)	13. (a)	14. (c)	15. (a)	16. (c)	17.(d)	18.(d)	19.(b)	20.(d)
21. (b)	22. (b)	23. (d)	24. (d)	25. (b)	26. (d)	27.(b)	28.(d)	29.(b)	30.(d)
31. (d)	32. (a)	33. (a)	34. (c)	35. (d)	36. (c)	37.(b)	38.(c)	39.(a)	40.(a)
41. (c)	42. (b)	43. (d)	44. (b)	45. (a)	46. (a)	47.(d)	48.(b)	49.(a)	50.(a)

MODEL TEST PAPER-6

1. 90% of the pungency in chilli is present in which part of fruits ?

(a) Seed (b) Placenta

(c) Pericarp (d) None

2. Highest hybrid varieties in vegetable are grown in

(a) Cabbage (b) Tomato

(c) Brinjal (d) Carrot

3. Which is not the variety of brinjal?

(a) Pant samrat (b) Pant bahar

(c) Pant rituraj (d) Pusa ankur

4. Storage temperature of Brinjal is

(a) 5-6°C (b) 7-8°C

(c) 10-11°C (d) 14-15°C

5. Which variety of tomato can be grown throughout the year ?

(a) Pusa Sadabahar (b) Pusa Ruby

(c) Sioux (d) None of these

6. 'Saponine' is present in

(a) Asparagus (b) Spinach

(c) Tomato (d) All of these

7. Which of the following brinjal varieties is resistant to aphid ?

(a) Annamalai (b) Manjari Gota

(c) Arka Sheel (d) All of these

8. 'Solasodine'(0.37 mg/100g) is present in

(a) Brinjal (b) Tomato

(c) Chilli (d) Potato

9. White brinjal is recommended for

(a) Diabetes (b) Cancer

(c) Skin problems (d) Heart disease

10. Fruit setting in brinjal usually occurs in the flower having

(a) Long and medium style (b) Medium style

(c) Short and medium style (d) Short style

11. The insect, which attacks potato both in field as well as in storage is

(a) Cut worm (b) Jassids

(c) Potato tuber moth (d) None

12. Potato contain how much % of protein

(a) 1.8 (b) 1.4

(c) 2.2 (d) 3.2

13. Earthing of potato crop is done

(a) 20 DAS (b) 40 DAS

(c) 50 DAS (d) None

14. Which of the following crop is stored in cold storage in India

(a) Potato (b) Onion

(c) Cabbage (d) Garlic

15. Segmental polyploidy is found in which vegetable crop?

(a) Potato (b) Beet root

(c) Radish (d) Onion

16. Which of the following used to control soil born pathogen in organic agriculture?

(a) Vermicompost (b) Trichoderma

(c) FYM (d) Oil cakes

17. Pollination in beetroot occurs through

(a) Water (b) Insects

(c) Both a & b (d) Wind

18. Potato is a

(a) Short day plant (b) Long day plant

(c) Day neutral plant (d) All of these

19. Inflorescence of beetroot is

(a) Spadix (b) Raceme

(c) Cymose (d) Spike

20. 'Forking' in carrot is due to ?

(a) Excess irrigation (b) Decaying organic manure

(c) Hard soil pan (d) Both b & c

21. Major sugar present in radish is

(a) Fructose (b) Sucrose

(c) Galactose (d) Glucose

22. **Seed rate of potato (tubers) is**
(a) 15-25 q/ha (b) 8-10 q/ha
(c) 40-45 q/ha (d) 35-40 q/ha

23. **Suberization temperature of potato is**
(a) 20°C (b) 25°C
(c) 30 °C (d) 35 °C

24. **Edible portion of radish roots develops from**
(a) Primary root (b) Hypocotyl
(c) Both a & b (d) None of these

25. **'Cavity spot' of carrot is due to**
(a) Balanced soil moisture (b) Ca deficiency
(c) Low nitrogen (d) All of the above

26. **'Bitterness' of carrot roots is due to the presence of**
(a) Low ethylene (b) High nitrogen
(c) High ethylene (d) High sulfur

27. **Red-violet colour of beetroot is due to the presence of**
(a) β – xanthins (b) β – cyanins
(c) Lycopene (d) None

28. **In cabbage Head to seed method is followed for the production of**
(a) Nucleus seed (b) Breeder seed
(c) Foundation seed (d) Both a & b

29. **Ancestor of cauliflower is**
(a) Brassica cretica (b) Brassica oleracea var. sylvestris
(c) Brassica oleracea var. rubra (d) None

30. **ETL of aphid population in potato is**
(a) 20/100 compound of leaves (b) 30/100 compound of leaves
(c) 40/100 compound of leaves (d) 10/100 compound of leaves

31. **Bacterial wilt resistant variety of chilli**
(a) Utkal Rashmi (b) Pusa Deepti
(c) Both a & b (d) Arka Basant

32. **Which variety of Okra is bred by Dr. Harbhajam Singh?**
(a) Pusa Mukhmali (b) Parbhani Kranti
(c) Punjab Padmani (d) Pusa Sawani

33. Which of the following is Shade loving crop?

(a) Ceylon spinach (b) Asparagus
(c) Sweet flag (d) Rhubarb

34. 'Yerusseri' is prepared from immatured fruits of

(a) Bottle gourd (b) Pumpkin
(c) Biiter gourd (d) None

35. 'Straight eight' is a variety of cucumber introduced from

(a) England (b) Australia
(c) USA (d) Australia

36. First gynoecious cucumber hybrid in India is

(a) Pusa Sanyog (b) Phule Subhangi
(c) Sheetal (d) Himangi

37. 'World Vegetable Centre' (WVC) is located at

(a) Geneva (b) Brazil
(c) Taiwan (d) USA

38. Which of the following cucurbit is amphidiploid nature?

(a) Snake gourd (b) Pumpkin
(c) Bitter gourd (d) Chow- chow

39. Summer squash is harvested at ... stage

(a) 1/24 maturity (b) 1/4 maturity
(c) 1/3 maturity (d) None

40. Ancestor of snake gourd is

(a) Trichosanthes origera (b) Trichosanthes dioca
(c) Trichosanthes kirilowii (d) Trichosanthes lobata

41. Seed rate of sponge gourd is

(a) 6-7 kg/ha (b) 2.5-3.5 kg/ha
(c) 7-8 kg/ha (d) 9-10 kg/ha

42. Pungency/flavour/bitterness in Cole crops is due to the presence of?.

(a) Solasodin (b) Oleoresin
(c) Sinigrin (glucosinolate) (d) Diosgenin

43. "Whip Tail' in cauliflower due to deficiency among which nutrient

(a) Bo (b) Mo
(c) Ca (d) Fe

44. Browning of curd associated with the deficiency of .

(a) Mo (b) Ca
(c) Bo (d) Zn

45. 'Sauerkraut' is a fermented product.

(a) Broccoli (b) Cauliflower
(c) Cabbage (d) Brussel sprout

46. Anti-cancer property of cabbage due to presence of chemical compound .

(a) Sinigrin (b) Indole-3-carbinol
(c) Glucosinolates (d) 2-3 tridicanone

47. The edible part of sprouting broccoli.

(a) Head (b) Curd
(c) Flower bud (d) Auxillary bud

48. Which among the following variety of broccoli has been developed from selection of exotic cultivar at IARI.

(a) Pusa broccoli KTS-1 (b) Palam Kanchan
(c) PalamVichitra (d) PalamSamridhi

49. Which variety of broccoli recommended for cultivation in the sub-tropical condition .

(a) Palam Kanchan (b) Palam Samaridhi
(c) PalamHaritika (d) PalamVichitra

50. The source of high glucoraphanin content is

(a) B. alboglabra (b) B. villosa
(c) B. cretica (d) B. nigra

Answer

1. (b)	2. (b)	3. (b)	4. (c)	5. (a)	6. (d)	7. (a)	8. (a)	9. (a)	10. (a)
11. (b)	12. (a)	13. (a)	14. (c)	15. (a)	16. (c)	17. (d)	18. (d)	19. (b)	20. (d)
21. (b)	22. (b)	23. (d)	24. (d)	25. (b)	26. (d)	27. (b)	28. (d)	29. (b)	30. (d)
31. (d)	32. (a)	33. (a)	34. (c)	35. (d)	36. (c)	37. (b)	38. (c)	39. (a)	40. (a)
41. (c)	42. (b)	43. (d)	44. (b)	45. (a)	46. (a)	47. (d)	48. (b)	49. (a)	50. (a)

MODEL TEST PAPER-7

1. **Pusa Jyoti is a variety of**
   (a) Amaranth (b) Bottle gourd
   (c) Spinach (d) Brinjal
2. **Which of the following is primary center for spinach ?**
   (a) Central Asia (b) Africa
   (c) Near East (d) Central America
3. **Pusa Swarnima is a variety of**
   (a) Summer Squash (b) Turnip
   (c) Winter Squash (d) Carrot
4. **Pusa Alankar is a variety of**
   (a) Summer Squash (b) Turnip
   (c) Winter Squash (d) Carrot
5. **In onion, the type of self sterility**
   (a) Genic (b) Cytoplasmic-genic
   (c) Cytoplasmic (d) None of the above
6. **Arka Harit is a variety of**
   (a) Amaranth (b) Bottle gourd
   (c) Bitter guard (d) Tomato
7. **Mitosis takes place in which of the following organs ?**
   (a) Reproductive organs (b) Somatic organs
   (c) Anthers (d) Both (a) and (b)
8. **Introduction of genes coding for Bacillus thuringenesis (BL) insecticidal crystal protein confer resistance to**
   (a) Insects (b) Bacteria
   (c) Diseases (d) Viruses
9. **Strains of Bt produce toxins, which confer resistance against**
   (a) Diptera (b) Coleopteran
   (c) Lepidoptera larvae (d) All the above
10. **Which of the following processing ? tomato varieties are good for**
    (a) Arka Shaurabh (b) Arka Ashish
    (c) Arka Ahuti (d) All

11. Colour blindness genes are located on

(a) Y-chromosomes (b) X-chromosomes

(c) Autosomes (d) All of the above

12. Which of the following is primary center of diversity for cucumber ?

(a) Chinese-Japanese (b) Hindustani Center

(c) Indo-Chinese (d) Mediterranean region

13. The vegetable in which all parts of plant can be used as edible part is

(a) Lima bean (b) Winged bean

(c) Radish (d) Broad bean

14. The capacity of a cell to develop into a complete plant is called.............

(a) Tissue culture (b) Fecundity

(c) Totipotency (d) Haploidy

15. Which of the following tomato variety has the indeterminate growth habit ?

(a) Sioux (b) Pusa Gaurav

(c) Roma (d) All

16. A pureline may be used

(a) As a parent for hybridization (b) For mutation studies

(c) As a variety (d) All the above

17. The orNamental type in case of cucurbits is

(a) Summer squash (b) White squash

(c) Pumpkin (d) White squash.

18. Herkogamy favours Self pollination

(a) Cross pollination (b) SP

(c) Mutation (d) Wide hybridization

19. Coloured onion var. are suitable for dehydration

(a) Red (b) Yellow

(c) White (d) Al

20. Fenugreek seed contain

(a) Allicin (b) Diosgenin

(c) Ajmacline (d) All of these

21. Pusa Ruby is a variety of

(a) Potato (b) Brinjal

(c) Tomato (d) Chilli

22. The single seeded cucurbit is

(a) L. acutangula (b) L. cylindrica

(c) C. moschata (d) S. edule

23. Onion is stored at a temperature of

(a) 4-5 °C (b) 25 °C

(c) 6-8 °C (d) 10 °C

24. Which is not a cytokinin ?

(a) BA (b) Zeatin

(c) Etheral (d) Kinatin

25. The book HORTI. SCIENCE is written by

(a) Willium gad (b) Swaminathan

(c) Jennick (d) Randhawa

26. Which of the following is responsible for controlling of fruit drops ?

(a) Auxin (b) Gibberellins

(c) Cytokinin (d) Ethylene

27. Self incompatibility results due to

(a) Genetic causes

(b) Morphological causes

(c) Physiological and Biochemical causes

(d) All of the above

28. Which is not a legume crop?

(a) Pea (b) beans

(c) Wheat (d) Fenugreek

29. In onion cross-pollination is encouraged by which mechanism?

(a) Protogyny (b) Protoandry

(c) Cleistogamy (d) Heterostyly

30. Genetic variability can be created through

(a) Somatic hybridization (b) Hybridization

(c) Mutation (d) All

31. Object of precooling is

(a) To fresh the product (b) To improve the quality

(c) Both a and b (d) To remove field heat

32. Sree Shilpa is first hybrid of

(a) Greater yam (b) Elephant foot yam

(c) Lesser yam (d) White yam

33. A cross in which order of male & female parent is reversed is called as

(a) Reciprocal cross (b) Back cross

(c) Test cross (d) Top cross

34. Horticultural Society of India (HST) was established in.......

(a) 1942 (b) 1947

(c) 1954 (d) 1950

35. Sree Harsha is a var. of

(a) Yam (b) Cassava

(c) Calocasia (d) Sweet Potato

36. Late blight resistant variety of Potato

(a) Kufri Jyoti (b) Kufri chipsona

(c) Kufri giriraj (d) All

37. Which one of the following vegetable produce maximum seeds per fruit ?

(a) Tomato (b) Brinjal

(c) Chilli (d) Potato

38. Anther culture is ordinarily used for production of

(a) Diploid plants (b) Haploid plants

(c) Polyploid plants (d) Triploid plants

39. Multistyled flowers are common in

(a) Brinjal (b) Okra

(c) Chilli (d) Limabean

40. Which of the following crop show protandry ?

(a) Carrot (b) Tomato

(c) Chilli (d) Brinjal

41. The 2n number of cucumber is

(a) 14 (b) 24

(c) 38 (d) 18

42. Daily requirement of vitamin D for a healthy adult is:

(a) 200 1.U. (b) 400 I.U.

(c) 600 1.U. (d) 800 I.U.

43. Botanical Name of mango ginger is

(a) Curcuma longa (b) Curcuma amada

(c) Mangifera indica (d) Curcuma aromatic

44. Plant exploration would be most successful when it is targeted at

(a) Farmer's field (b) Primary centre of origin

(c) A microcenter (d) Secondary centre of origin

45. The term epistasis is given by

(a) Mendel (b) Halman

(c) Bateson (d) Hull

46. Irish famine was caused by

(a) Phytopthora (b) Alternaria

(c) Fusarium oxysporum (d) None of the above

47. Which of the following insect in non polyphagous ?

(a) Lady bird beetle (b) White grub

(c) Termite (d) Grass hopper

48. Asexually propagated crops are

(a) Highly heterozygous (b) Generally polyploids

(c) Mostly perennials (d) all of these

49. Arka Suman is a variety of

(a) French bean (b) Cowpea

(c) Garden pea (d) Dolichus bean

50. Which one of the following is highly cross-pollinated?

(a) Okra (b) Bitter gourd

(c) Chilli (d) Cowpea]

Answer

1. (c)	2. (a)	3. (b)	4. (a)	5. (b)	6. (c)	7. (b)	8. (a)	9. (c)	10. (d)
11. (b)	12. (b)	13. (b)	14. (c)	15. (d)	16. (a)	17. (a)	18. (a)	19. (c)	20. (b)
21. (c)	22. (d)	23. (a)	24. (c)	25. (c)	26. (a)	27. (d)	28. (c)	29. (a)	30. (d)
31. (d)	32. (a)	33. (a)	34. (a)	35. (b)	36. (a)	37. (a)	38. (b)	39. (a)	40. (a)
41. (a)	42. (a)	43. (b)	44. (c)	45. (c)	46. (a)	47. (a)	48. (d)	49. (b)	50. (b)

MODEL TEST PAPER-8

1. **Maximum life of Vanilla is**
   (a) 10 years (b) 15 years
   (c) 30 years (d) 50 years
2. **Navashree and Nithyashree are varieties of**
   (a) Cardamom (b) Cinnamom
   (c) Black Pepper (d) None
3. **Commercial part of mace is**
   (a) Dried kernel (b) Seed
   (c) Flower (d) Dried aril
4. **Mudigree is a var. of**
   (a) B. Pepper (b) Cardamom
   (c) Cinnamon (d) Clove
5. **Saffron belongs to the family**
   (a) Orchidaceae (b) Iridaceae
   (c) Lauraceae (d) Apiaceae
6. **Eugenol is the main constituent in the leaf oil of ?**
   (a) Cassia (b) Cinnamon
   (c) Bal leaf (d) None
7. **Propagation of cardamom is done by**
   (a) Seeds (b) Sucker
   (c) Both a & b (d) Rhizome
8. **Pickings types of cardamom is**
   (a) Light picking (b) Hord picking
   (c) Both a & b (d) None
9. **Konkan Tej is the variety of**
   (a) Cardamom (b) Long
   (c) Cinnamon (d) Ginger
10. **Oil of Black pepper consist mainly of**
    (a) Phenol (b) Alkaloid
    (c) Terpense (d) Tetracycline

11. Alleppy and Duggirala are the trade Name of ......

(a) Ginger (b) Tamarind

(c) Ajowain (d) Cumin

12. 'Trachyspermum ammi' is the botanical Name of ?

(a) Hing (b) Ajowain

(c) Zeera (d) Coriander

13. Cinnamon belong to family

(a) Lauraceae (b) Fabaceae

(c) Iridaceae (d) Poaceae

14. Important ingradient in Bishop's weeds is

(a) Cineol (b) Thymol

(c) Cucumine (d) Sabinene

15. Family of Dill is

(a) Apiaceae (b) Leguminoacae

(c) Myrtaceae (d) Lauraceae

16. Garcinia and Kokam belong to the family of ?

(a) Apiaceae (b) Clausiaceae

(c) Myrtaceae (d) Lauraceae

17. Linalool, important ingradient present in

(a) Fennel (b) Cumin

(c) Coriander (d) Aniseed

18. Terbinen is the important constituent present in

(a) Sweet flag (b) Cinnamon

(c) Nutmeg (d) Cury leaf

19. Origin of small cardamom is

(a) India (b) Srilanka

(c) Indonasia (d) China

20. "Bareja system" is the term associated with

(a) Nutmeg (b) Betelvine

(c) All spice (d) Black Pepper

21. Which of the following is oldest known spice

(a) All spice (b) Nutmeg

(c) Cassia (d) Black pepper

22. Feathering and Scrapped chipps are important grades of

(a) Black pepper (b) Clove

(c) Cinnamon (d) Cardamom

23. The biting taste of Black pepper is ?

(a) Piperine (b) Pentanol

(c) Terpenine (d) Pipermenthene

24. Essential oil % in Dill/Sowa is

(a) 1.5-4 % (b) 0.1-1.0%

(c) 0.1-2.6% (d) None of these

25. Large cardamom is a

(a) Sun loving plant (b) Shade loving plants (seophyte)

(c) Both (d) None

26. Cinnamomum tamala is the botanical Name of:

(a) Cinammon (b) Indian cassia (Tejpat)

(c) All spice (d) None

27. Major all spice producing country in the world is

(a) Jamica (b) India

(c) USA (d) None of these

28. Curry leaf is a

(a) Self pollinated crop (b) Often cross pollinated crop

(c) Cross Pollinated crop (d) None of these

29. Sex form present in Betelvine is

(a) Monoecious (b) Dioecious

(c) Andromonecious (d) None of these

30. "Katte disease" main problem in which crop?

(a) Cinnamon (b) Cumin

(c) Cardamom (d) Pepper

31. Family of clove is

(a) Myrtaceae (b) Alliaceae

(c) Apiaceae (d) Lauraceae

32. World most expensive spice crop is

(a) Saffron (b) Nutmeg

(c) Turmeric (d) Clove

33. Piper nigrum is a botanical Name of

(a) Saffron (b) Black pepper

(c) Turmeric (d) Clove

34. Garcinia indica is a botanical Name of

(a) Kokun (b) Garcinia

(c) Turmeric (d) Clove

35. Cinnamon is which plant part is used?

(a) Bark (b) Seed

(c) Rhizome (d) Pods

36. Black pepper is which plant part is used?

(a) Bark (b) Seed

(c) Rhizome (d) Berries

37. Nutmeg is which plant part is used?

(a) Aril and Kernel (b) Seed

(c) Rhizome (d) Berries

38. Garcinia gummi-gutta is a botanical Name of

(a) Kokun (b) Garcinia

(c) Turmeric (d) Clove

39. 2n of Black pepper is?

(a) 52 (b) 48

(c) 63 (d) 22

40. 2n of Coriander is?

(a) 22 (b) 48

(c) 63 (d) 14

41. Origin of Cinnamon is

(a) India (b) Srilanka

(c) Indonasia (d) China

42. Origin of Nutmeg and Clove is

(a) India (b) Srilanka

(c) Indonasia (d) China

43. Clove is which plant part is used?

(a) Unopen flower bud (b) Seed

(c) Rhizome (d) Berries

44. Which crops is diocieous in nature?

(a) Nutmeg (b) All spice
(c) Juniper (d) All

45. Konkan Amrit is the variety of

(a) Kokun (b) Garcinia
(c) Turmeric (d) Clove

46. Suruchi is the variety of

(a) Ginger (b) Garcinia
(c) Turmeric (d) Clove

47. Family of Vanilla is

(a) Orchidiaceae (b) Alliaceae
(c) Apiaceae (d) Lauraceae

48. Funnel shaped stigma used as a spice in which crop?

(a) Kokun (b) Saffron
(c) Turmeric (d) Clove

49. Which crop only cultivated in Kashmir under rain condition is?

(a) Kokun (b) Saffron
(c) Turmeric (d) Clove

50. Vanilla is propagated by ?

(a) Stem cutting (b) Seed
(c) Fruits (d) Rhizome

Answer

1. (a)	2. (b)	3. (d)	4. (b)	5. (b)	6. (b)	7. (c)	8. (c)	9. (c)	10. (c)
11. (a)	12. (b)	13. (a)	14. (b)	15. (a)	16. (b)	17. (c)	18. (c)	19. (a)	20. (b)
21. (c)	22. (c)	23. (a)	24. (a)	25. (b)	26. (b)	27. (a)	28. (a)	29. (d)	30. (c)
31. (a)	32. (a)	33. (b)	34. (a)	35. (a)	36. (d)	37. (a)	38. (b)	39. (a)	40. (a)
41. (b)	42. (c)	43. (a)	44. (a)	45. (a)	46. (a)	47. (a)	48. (b)	49. (b)	50. (a)

Chemical Name and deficiency of vitamin

Vitamin	Chemical Name	Deficiency
Fat soluble		
Vit. A	Retinol	Dermatitis Xerophthalima Night blindness (Nyctalopia)
Vit. D	Calciferol -Ergocalciferol (D_2) -Cholecalciferol (D_3) -Alfacalcidol	Rickets Pigeon chest in children Osteomalacia in adults
Vit. E (Beauty vitamin)	Tocoferol	Sterility
Vit. K	Phylloquinone (K_1) Meraquinonoe (K_2)	Delayed blood clotting
Water soluble		
Vit. B_1	Thiamine	Bery-bery
Vit. B_2	Riboflavin	Chelosis
Vit. B_3	Niacin	Pellagra
Vit. B_5	Pantothenic acid	Mental disorder
Vit. B_6	Pyredoxin	Anemia
Vit. B_7/ Vit. -H	Biotin	Hair loss
Vit. B_9	Folic acid	Anemia
Vit. B_{12}	Cyanocobalamine	Anemia
Vit. C	Ascorbic acid	Scurvy

CHAPTER-24

Matching Question

1. **Matched the following**

(a) Pusa Madhavi	(i) Onion
(b) Pant Anupama	(ii) French bean
(c) Hisar Lalit	(iii) Tomato
(d) CH-1	(iv) Chilli
(e) Pusa Kesar	(v) Carrot

Ans. a-i, b-ii, c-iii, d-iv, e-v

2. **Matched the following**

(a) Sweet potato	(i) 90
(b) Potato	(ii) 48
(c) Brinjal	(iii) 24
(d) Pea	(iv) 14
(e) Cabbage	(v) 18

Ans. a-i, b-ii, c-iii, d-iv, e-v

3. **Matched the following**

(a) Napiform	(i) Sweet potato
(b) Conical root	(ii) Radish
© Fusiform root	(iii) Tomato
(d) Adventious root	(iv) Suagrbeet
(e) Simple root	(v) Carrot

Ans. a-iv, b-v, c-ii, d-i, e-iii

4. **Matched the following**

(a) Swarna Rekha	(i) Okra
(b) Great Lakes	(ii) Pointed gourd
© Pusa Parvati	(iii) Crips head

(d) Pusa Shubhra	(iv) French bean
(e) Arka Abbay	(v) Cauliflower

Ans. a-ii, b-iii, c-iv, d-v, e-i

5. Matched the following

(a) Rhubarb	(i) Lye peeling
(b) SO_2 indicator	(ii) Siberia
(c) Hottest chilli	(iii) Lichen
(d) Chlorophyll	(iv) Bhut Jolokia
(e) NaOH	(v) N_2

Ans. a-ii, b-iii, c-iv, d-v, e-i

6. Matched the following

(a) Hollow heart	(i) White fly
(b) Iodine	(ii) Potato
(c) Blanching	(iii) Okra
(d) BER	(iv) Inactive the enzyme
(e) Leaf curl virus	(v) Physiological disorder

Ans: a-ii, b-iii, c-iv, d-v, e-i

7. Matched the following

(a) Pusa Naveen	(i) Placenta
(b) Ethylene	(ii) Without crook neck
© Transfer disease resistance	(iii) Methionene
(d) Sabudana/Sago	(iv) Back cross
(e) Chilli	(v) Cassava

Ans. a-ii, b-iii, c-iv, d-v, e-i

8. Matched the following

(a) BER	(i) Biodiversity International
(b Natural harmone	(ii) Ca
© Ethylene	(iii) IAA
(d) Crossing over	(iv) Methionine
(e) IBPGR	(v) Pachytene

Ans. a-ii, b-iii, c-iv, d-v, e-i

9. Matched the following

(a) Black heart	(i) Potato
(b) Golden fleck	(ii) Tomato
(c) Pillow	(iii) Cucumber
(d) Whiptail	(iv) Cauliflower
(e) Pithiness	(v) Radish

Ans. a i, b ii., c iii, d-iv., e-v

10. Matched the following

(a) CIMAP	(i) Lucknow
(b) IIHR	(ii) Banglore
© CSIR	(iii) New Delhi
(d) CTCRI	(iv) Thiruvanantpuram
(e) CFTRI	(v) Mysore

Ans. a-i, b-ii, c-iii, d-iv, e-v

11. Matched the following

(a) Bud killing	(i) Excessive N_2
(b) Interveinal chlorosis	(ii) Mg deficiency
© Leaf scorching	(iii) Cl toxicity
(d) BER	(iv) Ca deficiency
(e) Fruit cracking	(v) Bo deficiency

Ans. a-i, b-ii, c-iii, d-iv, e-v

CHAPTER-25

Confusing Varieties of Different Crops

Variety	Crop	Variety	Crop
Punjab Sadabahar	Brinjal, Chilli	Pusa Bindu	Brinjal
Pusa Sadabahar	Tomato, Chilli, Cluster bean	Arka Bindu	Onion
Arka Sumeet	Muskmelon, Ridge gourd	Arka Komal	French bean
Azad Kranti	Brinjal, Okra	Pusa Komal	Cowpea
Arka Suryamukhi	Winter squash	Punjab Komal	Bottle gourd
Phule Suryamukhi	Chilli	Pusa Uphar	Tomato
Phule Jyoti	Chilli	Pusa Upkar	Brinjal
Pusa Jyoti	Palak	Pant Uphar	Pea
Arka Meghali	Tomato	Annamalai	Muskmelon, Brinjal
Pusa Meghali	Carrot	TA-19	Snake gourd
Pusa Harit	Palak	T-19	Pea
Arka Harit	Bitter gourd	Pant Bahar	Tomato
Hisar Harit	Pea	Arka Bahar	Bottle gourd
Pusa Ageti	Cabbage	Punjab Bahar	Brinjal
Punjab Ageti	Radish	Pusa Sheetal	Tomato
Pant Anupama	French bean	Arka Sheetal	Long melon
Arka Anupama	Palak	Pusa Pragati	Pea
Jade	Okra	Arka Pragati	Onion
Jade Cross	Brusseles sprouts	Hisaar Pragati	Brinjal
Pusa Sunheri	Sweet Potato	Pusa Safed	Sweet Potato
Punjab Sunheri	Muskmelon	Punjab Safed	Radish
Konkon Tara	Bitter gourd	Marvel	Tomato
Konkon Harita	Ridge gourd	Little Marvel	Pea
Konkon Sweta	Snake gourd		
Harita	Sponge gourd		

9 789359 199603

Printed at Repro India Ltd.